ELECTROMAGNETIC
FIELDS

B. BLAKE LEVITT

ELECTROMAGNETIC
FIELDS

A Consumer's Guide
to the Issues and
How to Protect Ourselves

AN AUTHORS GUILD BACKINPRINT.COM EDITION

Electromagnetic Fields
A Consumer's Guide to the Issues and How to Protect Ourselves

AN AUTHORS GUILD BACKINPRINT.COM EDITION
Published by iUniverse, Inc.

For information address:
iUniverse, Inc.
2021 Pine Lake Road, Suite 100
Lincoln, NE 68512
www.iuniverse.com

Originally published by Harcourt Brace/Harvest Books

Because of the dynamic nature of the Internet, any Web addresses
or links contained in this book may have changed
since publication and may no longer be valid.

The views expressed in this work are solely those of the author and do not
necessarily reflect the views of the publisher, and the publisher
hereby disclaims any responsibility for them.

Illustrations by Tamara L.Newnam

ISBN: 978-0-595-47607-7

Printed in the United States of America

This book is dedicated to all those
who have gone before me on this subject
and to
Jon Garvey, my husband and pal.

ACKNOWLEDGMENTS

T HE SUBJECT OF electromagnetic fields has fascinated me for nearly two decades. At first glance, it might seem like any other environmental pollutant (albeit an esoteric one) that could contribute to health problems in humans. But in fact, the implications are immense and go directly to the heart of twentieth-century life. The medical underpinnings alone may prove to be among the most significant anatomical understandings that science has reached to date.

There is a reason why so few medical and science journalists have gotten their teeth into this subject. To write thoroughly about electromagnetic fields, one has to delve into physics and electrical engineering, geology and the earth sciences, biology, botany, history and medical anthropology, anatomy and physiology, the veterinary sciences, and cellular and molecular biology, as well as political science and public policy. Many journalists have been pulled to one area or another, but few have taken on the totality of the subject. In attempting to do just that, I have drawn on the impeccable research, writing, and dedication of many others. First and foremost, I would like to acknowledge some of them.

Thanks are in order to Dr. Louis Slesin, editor and publisher of *Microwave News,* who has been the "Ralph Nader" of the subject for going on twenty years—holding everyone's feet to the fire, both pro and con. And to Paul Brodeur, a fire-and-brimstone journalist (whom some would call a pit bull for facts), who has kept the lay reader's attention focused on the politics of the subject since the 1970s. And especially to Dr. Robert Becker, who has organized our thinking on electromagnetic fields and the products all around us in a way that will forever change our perspective. Dr. Becker is a true researcher in the best sense of the word, embodying the wonder and clarity of spirit to witness the way things unfold as they do—then to wonder some more. And also to Dr. Andrew Marino, a colleague of and co-researcher with Dr. Becker, who understood this not only as a pioneering medical issue but also as an eventual legal one, and so became a doctor of jurisprudence as well as a doctor of biophysics—in addition to writing his own seminal books on the subject. And to the other pioneers whose work I have drawn from: Dr. W. Ross Adey, Suzanne Bawin, Dr. Carl Blackman, Dr. David O. Carpenter, Dr. José Delgado, Dr. Reba Goodman, Ed Leeper, Dr. Abraham Liboff, Dr. Rochelle G. Medici, Dr. David Savitz, Dr. Cyril Smith, and Dr. Nancy Wertheimer, as well as the scores of others whose work I have cited. Acknowledgment is also in order to those in various government agencies who have taken a pro-active position even in the face of political pressure.

Recognition is due to writers before me on the subject, too: Roger Coghill, Simon Best, Gary Selden, Ellen Sugarman, Kathleen McAuliffe, Peter Tompkins, and Christopher Bird, among others.

For their patience in proofreading specific chapters, special thanks are in order to my content editors: Louis Slesin for initial feedback; Robert Becker for fine-tuning my medical understanding as well as for important encouragement; Karl Riley of Magnetic Sciences International, a Teacher (with a capital T) who not only fixed my technical errors but took the time to make sure I understood why I had made them; Carroll Adam Cobbs, who helped immeasurably with the research and editing of the radio-frequency sections; and Robert Carberry of Northeast Utilities, who assisted with research, perspective, and the proofreading of some of the illustrations. And special thanks to Dr. Don Deno (an engineer's engineer if ever there was one) and Dave Murray of Electric Field Measurements, for letting me pick their stellar brains for several hours.

In addition, acknowledgments are certainly in order to a host of activists across the country, who seemed to spring up quickly, to network intensively, and to share information more willingly than other groups I've known. Those who generously sent me information include John Bogucz, a networker par excellence; Dr. Robert Dudney, a fine physician who embodies the kind of spiritual vision that so many clinicians unfortunately have lost; Nancy Weidinger, a grandmother with impeccable instincts and a citizen's library on electromagnetic fields to rival the Library of Congress; and Kathy and Robert Hawk, whose tenaciousness has served to educate an entire state about cellular-phone towers. And to the countless others I have yet to meet who have taken the time to educate themselves on a very difficult subject.

And to those who provided their fine make-it-happen skills, without which no book would ever come into being: to John Radziewicz of Harcourt Brace, who had the courage to take on a book for the consumer well before the recent media attention aroused public interest in the subject; and to my editor, Vicki Austin-Smith, who truly knows how to let writers write while encouraging them each step of the way. And to my illustrator, Tamara Newnam, who understood things better over the phone than most people ever do in person. And to Carole Abel, the best agent one could find — the kind an author comes to call friend.

Thanks, too, go out to friends Merle Bombardieri, Jean Hoegger, and Jack Haig, who were among the first to encourage me, as well as to Charlotte Libov and Terry Montlick — just because.

And, perhaps most importantly, gratitude is in order to my family: to my husband, Jon Garvey, who proofed the chapters, relieved me of many domestic obligations, ran interference on a number of levels, and put up with my chaos throughout this long project. And to my mom and all the other family members who understood, offered encouragement, and gave me the necessary space to accomplish what I set out to.

CONTENTS

PART ONE
ELECTROMAGNETIC FIELDS: A COMPLEX SUBJECT

FOREWORD

THIS BOOK, first published in 1995 by Harcourt Brace/Harvest Books, is still considered one of the classics in the genre. No other book on the health/environmental effects of non-ionizing radiation covers such a broad range of information for the lay-reader and professional alike, making an arcane area of science accessible in plain language, without over-simplification. Everything in the book remains current and significant today. While cell phones and towers, broadcast facilities and many wireless products are covered thoroughly in the book, we have seen a veritable juggernaut of wireless devices appear within the last decade. Unfortunately, none of the questions regarding the safety of these devices has been settled, all the while more research points to the need for caution. The safety of such exposures becomes more critical with each passing day. There is almost no such thing as a non-exposed population now.

The first edition of the book appeared on the eve of the Telecommunications Act of 1996—a bill that would forever change the way we communicate, as well as our radiofrequency (RF) regulatory landscape. The Telecom Act contained an insidious clause, inserted at the behest of the telecom industry, which forbids communities from taking the environmental effects of RF into consideration when siting cell towers. (Though the language in the bill specifies "environmental effects," it is widely interpreted to mean health effects in humans. This preemption now applies to cell towers and broadband services, but not broadcast or other RF-transmitting facilities.) Key aspects of this federal usurpation of local public health authority had never been done before. It was a massive preemption of health regulatory oversight given to a non-health agency—the Federal Communications Commission (FCC)—to adopt/enforce standards for radiation emissions. At the time this was being enacted, the federal agency with actual oversight for such environmental radiation exposures—the U.S. Environmental Protection Agency (EPA)—had its bioelectromagnetics research program slashed. Because of that preemption, many communities came to think that RF "couldn't be talked about" at municipal meetings. Citizens who raise health concerns today about cell towers are often summarily shut down by their planning/zoning boards. But since when can't we talk about something in America? That's a First Amendment violation of free speech.

Quite simply, modern society is in the throws of a massive global experiment with a new energetic form of air pollution, called electrosmog or electropollution. It's biophysics plain and simple—the area where all our wireless gizmos created by the physics and engineering disciplines meet with the biological systems in their paths. Electrosmog has now been found capable of effecting the DNA of every living thing and it may even be having an adverse impact on the earth's atmosphere. Electrosmog will likely prove to be the greatest environmental challenge of this century. We appear to be on a direct collision course with the basic building blocks of life.

Many people today think that concerns over powerline electromagnetic fields in the extremely low frequency (EMF/ELF) range have been settled and given a clean bill of health, but nothing could be farther from the truth. Three U.S. state and federal agencies now recognize that such exposures are directly linked with childhood leukemia. So do several European governments and the European Union. New studies indicate that living near powerlines during childhood significantly increases the lifetime risk of developing adult leukemia as well. No one had followed that population so far into the future until recently. But as is so often the case with this subject, as soon as a researcher looks, effects are found. The states of California and Connecticut are establishing guidelines for EMF/ELF exposures, in some cases requiring new powerline corridors to be placed underground near populated areas. The UK is considering buying residential properties along powerline corridors to create larger setbacks for the population. Plus, there are new concerns by health physicists over something called "dirty electricity," a term used to describe the increasing multi-frequency exposures created when RF from electronic devices couples with powerlines and plumbing fixtures, following their pathways into our homes and businesses. Dirty electricity has been found to cause increases in diabetes, headaches, concentration problems, and hyperactivity in school children, among other problems.

But the greatest concern regarding electrosmog today is over wireless products. Since the advent of cell phones in the late 1980s, the "wireless revolution" has grown to include everything from wireless Internet access in schools, libraries, homes, airports and hospitality businesses, to Wi-Fi "hotspots" anywhere people gather—from Starbucks to McDonald's to the local food co-op—as well as numerous remote sensing devices and radiofrequency identification (RFID) tags now imbedded in many consumer products, passports, dementia patients, pets, and children in neonatal units. People are wearing Bluetooth "RF earrings" that pulse RF directly into their skulls. WiMax is a personal wireless network that allows people to bundle all of their communication services together thereby going completely wireless. The number of people abandoning their landline phones altogether grows daily. At the behest of the telecommunications giants again, the FCC in 2006 wrapped all new broadband services—requiring a lot more ambient RF—under the preemptions of the Telecom Act, which still forbids ambient RF exposures from being taken into consideration in tower siting. Add to this many new military applications, such as 'Star Wars' missile defense shield programs that create intense screens of high powered energy in the ionosphere, increased exposures to our military men and women from high-tech warfare, and hundreds of new radio and TV channels broadcasting at literally millions of watts of effective radiated power from both ground-based transmitters and satellite networks and it is clear we are awash in ever-increasing levels of ambient RF radiation.

Much of this new infusion of environmental RF is in the ultra high frequency (UHF) microwave (MW) bands—the same form of energy used by microwave ovens but without walls or doors. Not only are we being exposed from our own products but, like secondhand smoke, from those used by others too. Wireless consumer products transmit in a 360-degree pattern, exposing everyone in both the near and far field. Children today, for the first time in human history, have prenatal-to-grave exposures. Parents place baby monitors that transmit RF next to infant's cribs to keep them "safe." And children, at least in America, now get cell phones long before they get a driver's license. Parents often insist that children carry cell phones. How ironic that science is increasingly sounding alarm bells but parents remain largely uninformed.

There are over 2.2 billion cell phone users in the world today with an estimated 100,000 more signing up daily in America alone. Plus, every new technological offering, such as digital pictures sent via cell phones, wireless music downloads and text messaging, adds another layer of RF. All of this requires transmission infrastructure and increasing bandwidth in the electromagnetic spectrum, which means more transmitters in close proximity to the population. Just a decade ago, these exposures barely existed. One estimate holds that ambient background levels of RF in Boston and New York City have increased 100-fold over a decade ago. A recent study done in Europe found a 3000 percent increase in ambient RF in 10 cities versus a decade ago. By anyone's reckoning, that's a changed environment. There are virtually no other parallels today with any known pollutant.

Prior to the 20th Century, the earth's natural electromagnetic background consisted of the north and south poles, the earth's own steady magnetic fields that emanate from its molten core, and lightening from thunder storms. Very little RF, and even less microwave radiation, reached the earth from space. Now that entire area of the electromagnetic spectrum has been filled in at the earth's surface, due exclusively to our own electropollution within that last 50 years. Plus, we have created signaling/propagation characteristics like alternating current and digital pulsing, and unusual wave forms such as sine and sawtooth waves, and high power densities in these bands that simply do not exist in nature. These are all man-made artifacts.

Many public health professionals in Europe now recommend prudent avoidance of these exposures, especially for children who, by virtue of their age, are in a higher state of cell division, have thinner skulls that allow for deeper energy penetration, and have developing immune systems. But new technologies appear at breakneck speed. People seem addicted to their BlackBerrys and instant messaging systems, and they may well be. Some studies have found that RF increases endorphin levels and stimulates pleasure centers in the brain. Manufacturers have not warned users that such devices produce strong ELFs as well as an RF signal that is typically aimed directly at people's eyes. Two recent studies have found melanoma of the eye with such exposures. One study found ELFs at 900-milligauss levels with some BlackBerrys. (By comparison, increases in leukemia rates begin at 2 milligauss.)

In addition, whole cities are going Wi-Fi without thoroughly researching the subject. New Orleans, after Hurricane Katrina, may never fully rebuild a landline phone system again. Antenna arrays are mounted on towers, atop tall buildings and water tanks, clipped to the sides of apartment complexes, and regularly hidden in elevator shafts, barn silos and church steeples. In cities they are mounted on lampposts, creating a ubiquitous exposure within 10 feet of most city streets and apartments, directly irradiating people 24/7. No city-wide Wi-Fi system should be activated without epidemiology studies, too.

Citizens are fighting such systems in San Francisco, several southern California towns, Minnesota, Massachusetts, Maine,

Illinois, and throughout Canada and Europe. People are being made sick once systems go online. The German government in 2007 warned its citizens against Wi-Fi technologies and wireless routers, especially in schools. They advise that people keep computers and phones wired. Even cordless domestic phones are not recommended. In the US, many of these technologies are granted "categorical exclusions" and are unlicensed and unregulated by any government entity because they are presumed to operate below a certain power threshold for tissue heating. But there is no government oversight or monitoring and independent labs have often found such devices out of compliance with the few voluntary standards that exist.

Numerous communities today are blocking communications towers near schools and residential neighborhoods. But hundreds of thousands of new cell towers are needed to go online due to demand for all-things-wireless. The build-out of the wireless infrastructure is nowhere near complete. The "next generation" technology needs infrastructure every 1–3 miles apart. These are the new broadband services that allow everything from cell phone calls, text messaging and Internet downloads onto a cell phone screen or wireless laptop. Plus, in the name of competition with the cable networks, the telecoms can now offer TV programming. This means that yesterday's cell tower has morphed into today's broadcast facility. As a consumer, if you want these wireless services, then you are asking someone to live near the infrastructure. Eventually, it will be your backyard too. There's no way around it.

Science is also revealing a disturbing picture of what repeated RF exposure may be doing to challenge the health of our planet at all levels—from plant to human. There is a growing body of evidence that long-term, low-level exposure to RF is every bit as dangerous as short-term exposures like that from x-rays. We are, in essence, dealing with a whole new form of pollution—an energetic form. The non-human world is affected too. Electromagnetic radiation can cause trees to lose leaves prematurely and become more susceptible to diseases. Evidence shows that RF from cellular, TV, and radio towers lowers milk production in cows, causes deformities in amphibians, lowers reproduction in animals and birds, and causes confusion, navigational disruption and death in migratory birds. Bees' navigational abilities are known to be sensitive to low-level EMFs. The U.S. Fish and Wildlife Service now offers a conservative estimate that between 4-to-5 million bird deaths per year may result from bird collisions with towers. The songbird populations of industrialized countries are plummeting for myriad reasons. But RF may play a role as an attractant to birds since their eyes, beaks, and brain tissue are loaded with magnetite, a natural mineral highly sensitive to external magnetic fields that birds use in navigation. Noted American ornithologist, Robert Beason, discovered rapid neuronal firings in avian brain tissue exposed to cell-frequency RFs at very low intensities. And there are also indications that RF may be contributing to global warming through the agitation of hydrogen molecules in the upper atmosphere and ionosphere, like a microwave oven agitates water molecules in a coffee cup. Maybe greenhouse gases are not the only culprit in global warming. RF may prove a significant but hidden factor, according to some research.

Our best and brightest are largely unaware of this area of science. Environmentalists often attach radio collars and insert RFID tags in animals, birds, and marine populations in order to track and study them—without having read the literature that shows increases in tumors in living tissue with such devices as well as increased excitability and morbidity. Environmentalists have also embraced the replacement of incandescent light bulbs with compact fluorescents lights (CFLs) without understanding that CFLs create high magnetic fields up to 500 milligauss at the base of the bulb, often near a user's head, and also contribute to dirty electricity. Light emitting diode (LED) bulbs are far cleaner, emit no EMFs, use less energy and last three times longer than CFLs.

And many progressives have embraced free Wi-Fi Internet access as the answer to the "digital divide" between rich and poor communities—without understanding the potential health consequences to communities from both the wireless hardware as well as the infrastructure. Progressives have also embraced the concept of low-powered (1000 watt) FM radio stations to help bring local information to communities that have been disenfranchised by media consolidation. But this solution, which will bring a new layer of electrosmog to local areas, is advocated without understanding the research that has found increases in melanomas from low-power FM radiation. A better solution would be de-consolidation. Clearly, there are large gaps in education here.

Far more caution is warranted. Voluminous science exists for people who look. Environmentalists need to wake up to this new far-reaching threat to the planet. The progressive community needs more breadth in its research. Planning and zoning commissions need not be intimidated by the Telecom Act and should enact strict tower siting regulations. Citizens must hold their legislators accountable for creating appropriate, unbiased government research, truly protective standards, and to keep industry influence at arms-length from regulatory agencies. The science press needs to cover this subject as a medical issue, not just a wonders-of-the-technology story. And our medical clinicians need to comprehend what some biophysicists have long known—that living systems are exquisitely sensitive to low-level non-ionizing radiation. MD's have been traditionally taught in medical school that this kind of radiation can only cause tissue heating and electrical shock. But in fact, it can impact every system of the body.

The handwriting is on the wall. We need only open our eyes to make safer purchasing decisions for ourselves, our families, our planet. The adage "If it seems too good to be true, then it probably is ..." may come to haunt us. We may want to rethink many of these "magic" technologies. If it's wireless, we may well need to worry.

B. Blake Levitt, Connecticut, 2007

Other books by B. Blake Levitt:
(For ordering information, www.blakelevitt.com)

. BEFORE YOU CONCEIVE, THE COMPLETE PREPREG-
NANCY GUIDE, (Bantam, 1989) with John. R. Sussman, MD.

. 50 Essential Things To Do When The Doctor Says It's Infertility,
(Plume, 1995; iUniverse edition, 2000)

. Cell Towers, Wireless Convenience? Or Environmental Hazard?
(Safe Goods/New Century Publishing, 2001)

For more information:

. www.bioinitiative.org
A new report issued by an international group of scientists saying
that the current standards are not protective of the public health.
Up-to-date science and policy recommendations by some of the
foremost researchers in bioelectromagnetics.

. www.emf-portal.de/index.php
For scientific abstracts of studies both pro and con

. www.iaff.org/hs/facts/celltowerfinal.asp and
  www.iaff.org/hs/resi/celltowerfinal.htm
A position paper/resolution by the International Association of
Firefighters (IAFF)—the world's largest union of first responders—
prohibiting cell towers from being installed on firehouses until
more research is done. Firefighters are suffering documented
neurological damage from such installations.

. www.emrpolicy.org

Recent case law filed since The Telecommunications Act of 1996 was enacted. Also health/environmental studies and *amicus* briefs filed in lawsuits challenging the FCC. Look for a well-researched *amicus* brief filed by the Healthy Schools Network, which recommends no cell towers near schools. Also look for an *amicus* brief filed by the International Association of Firefighters (IAFF) which further describes their prohibiting cell towers from installation on firehouses. At this site, also look for the <u>FCC Flaws Fact Sheet</u> by B. Blake Levitt for information on why the standards are not protective of public health.

. www.emrpolicy.org/litigation/case_law/docs/exhibit_a.pdf

Questions raised in 1999 by the Radio Frequency Inter-Agency Work Group (RFIAWG)—a panel of federal agencies with a stake in RF issues—that have yet to be answered.

. www.microwavenews.com

Commentary, analysis, research and lots of information on this issue. Published by Louis Slesin, Ph.D.

. www.energyfields.org

Council on Wireless Technology Impacts—for professional resolutions and other information.

. http://www.c-a-r-e.org/

Canyon Area Residents for the Environment—Excellent site for broadcast issues.

. www.lessemf.com

For measuring and mitigation products, as well as books.

. www.MapCruzin.com

For maps of tower locations in your community and throughout the country.

Radiofrequency Antennas and Towers Across the U.S.

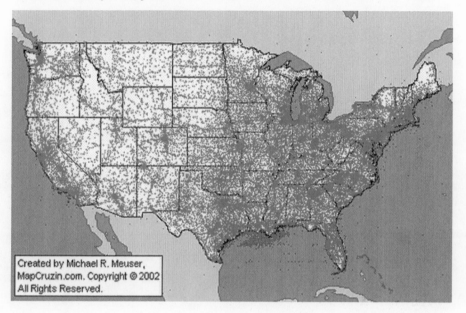

ELECTROMAGNETIC FIELDS: A COMPLEX SUBJECT

Chapter 1

HOW CAN THIS BE?

A Normal Twentieth-Century Day

IN ALL PROBABILITY, you were awakened this morning by the sound of an electric alarm clock or clock-radio placed within a foot of your bed. You slept last night with your head close to a wall that undoubtedly has electrical wiring running through it. After you got up, you may have made coffee on an electric stove or in a plug-in coffee machine. Your breakfast perhaps jumped from a toaster or was heated in a microwave oven, was poured from a blender or extracted from a juicer. You may have grabbed a quick look at the morning news on television or gone for a jog along your street, which is intersected by utility lines. After a shower, you probably used an electric razor, an electric toothbrush, or a hand-held hair dryer, and maybe you made a call on a cordless phone.

Perhaps you commuted to work on an electric train. If you drive to work, you may have opened and closed your garage door with a remote control. Possibly the entire stretch of highway you traveled ran parallel to a major utility corridor, or you were commuting side by side with drivers using mobile phones, even if you were not using one yourself. Just before you reached your office, maybe you stopped briefly at a store for a newspaper or a muffin and paid for it at one of those hip-level checkout scanners that automatically read the bar codes on packages. Then perhaps you walked through a number of doors that opened automatically as you approached.

Once inside your office, the first thing you did was flip on the overhead lights, which likely are fluorescent; then you turned on your computer screen and fired up the fax machine, the copier, and the printer. If you cared to investigate, you might have found a satellite dish or radio transmitters operating on top of the building where you work or live. Or, if such communications equipment isn't on your own building, you may have fallen within the umbrella of the antenna pattern from equipment located on an adjacent building or ridge line. Perhaps there is an electric substation in your neighborhood, either aboveground and therefore visible, or somewhat obscured from your sight inside a windowless brick or metal structure, or belowground—as is common in metropolitan areas—and therefore you are unaware of its presence altogether.

After work, you probably repeated most of these activities in reverse order, adding more time in front of a microwave oven or an electric stove and a few hours of TV watching before going to bed.

Congratulations. Like millions of others, you have been exposed to a constant electromagnetic field (EMF) of various frequencies and intensities, continuing all day, every day, and probably all night. And, like millions, you probably have no clear understanding, and no government guidelines, to alert you to the potentially serious bioeffects on you and your family. In fact, just living in a major metropolitan region will increase your exposure to ambient background levels of electromagnetic fields (EMFs) at least threefold, by some estimates, over that of those living in suburban or rural environments, although the degree of any exposure is dependent on your proximity to whatever source is creating it.

Serious Concerns

The human race has never before in its evolutionary history been exposed to such fields on a continuous basis, and there are serious and mounting concerns about the effects not just on individuals but on our entire ecosystem. Since the turn of the century, and increasingly since the early 1940s with the development of radar during World War II and the rapid growth of radio and TV broadcasting, we have surrounded ourselves with a veritable sea of artificially produced electromagnetic fields, all with a presumption of safety that many now think should never have been made.

Researchers have found a steady increase in many forms of cancer (independent of cancers with smoking or population growth factors) that some think closely parallels the increase in ambient electromagnetic fields, which includes radio frequencies and microwaves (RF/MW). A few researchers speculate that glandular and central-nervous-system cancers in particular are so related. What's more, although their findings are still largely speculative, some scientists are finding correlations between exposure to electromagnetic radiation and immune-system disorders like chronic fatigue immunodysfunction syndrome, HIV infections, and what some now refer to as electromagnetic sensitivity syndrome—a constellation of symptoms that include an inability to concentrate, mild panic, disorientation, visual disturbances, skin rashes, muscle weakness, and feelings of faintness. Not surprisingly, the syndrome is reported in computer operators, electronics workers, pilots, surgeons, and operating-room personnel more frequently than in other professions—that is, among people who work in continuous, overlapping electromagnetic environments, the effects of which have never been studied.

Consider chronic fatigue immunodysfunction syndrome— cavalierly scorned as the yuppie flu. Over the last ten years, what segment of the population is more likely to have been in a constantly elevated electromagnetic environment than these young urban professionals, with their high discretionary incomes? They are the very people most regularly using computers, cooking in high-tech kitchens, talking on cordless or cellular phones, and relaxing in entertainment rooms featuring multi-VCRs, big-screen TVs, and an assortment of remote-control devices, all of which emit various kinds of EMFs.

Or take the case of veterans of the Persian Gulf War, who were exposed to a barrage of high intensity EMFs unprecedented in military history and are reporting cases of chronic fatigue and immunosuppression in record numbers. The disorders were originally attributed to their exposure to toxic petroleum from the burning oil fields, but some now think that this alone does not explain the debilitating syndrome that afflicts so many of these veterans.

The Global Environment

The environmental impact of electromagnetic fields may well be altering our world in ways we do not understand. Some species of dolphins and seals are manifesting immunodeficiency diseases. In addition, whole stands of forests are dying, and frogs are disappearing all over the world at an alarming rate.

An early pioneer in EMF research, Elmer Lund, at the University of Texas, found back in the 1920s that frogs' eggs could be affected by both electric and magnetic fields; and Harold Saxon Burr, at Yale University, hooked up voltmeters to trees and discovered that their fields varied in response to external electrical events like thunderstorms, sunspots, and different phases of the moon, as well as to light and moisture. Today we may be seeing writ large the dire results hinted at by these early studies.

Some European scientists theorize that the devastating deforestation in the German, Austrian, and Swiss mountains is not solely due to acid rain, as was previously thought, but rather to the convergence of complex electromagnetic fields from high-powered communications transmitters for TV, radio, cellular phones, and radar, especially in the microwave bands. EMFs emitted by an abundance of information-gathering devices left over from the cold war, coming both from western Europe and from the former Soviet-bloc countries, converge in this region.

German scientists had long suspected that an unknown co-factor was involved in the deforestation. Recent studies hypothe-size that the configuration of pine needles and the leaf veins of deciduous trees make them perfect antenna receivers, with their hydrogen molecules acting as an electrical conductor for EMFs at different frequencies. Also, it is thought that microwave fre-quencies may cause the cell membranes of trees to resonate and thereby interrupt the tree's water circulation.

Microwaves are also thought to act on soil, plant life, and water in much the same way as microwave ovens act on food, only slower. Minute electrical currents travel down a tree trunk, flow into the ground in the form of electrical current, enter the soil, and create a type of acidification by means of an electrolytic pro-cess that effectively kills all living microorganisms there, leaving the soil unable to sustain new growth.

In the United States, tree damage can readily be observed along ridge lines, where it is common for radio transmitters to bounce their signals. Farmers have also observed for many years that grass often doesn't grow under an electric fence. But U.S. scientists have not yet taken as much interest in the subject as the Europeans have, and EMFs are not an important part of the American environmental dialogue, beyond their impact on the human body.

This may soon change, due in part to scientific curiosity about the disappearance of many species of frogs all over the world within the last three to five years. Whole aquatic communi-ties have disappeared in pristine areas in the course of one mating season, even where the water has been found to have a normal Ph balance and to be free of toxins. It is a genuine enigma that does not bode well for the planet. Some speculate that the mysterious disappearance may be due to increased ultraviolet radiation as a consequence of the shrinking ozone layer. But it would be interesting to investigate the relationship of electromagnetic fields a little lower on the spectrum than the ultraviolet band, given Dr. Lund's findings of so many years ago.

EMFs and the Human Anatomy: The Cart before the Horse

To say that we have put the cart before the horse would be a mild statement regarding the subject of artificially created electromagnetic fields and the human anatomy.

The cutting edge of research in Western medicine is only now coming to recognize the human body as a fundamentally coherent electrical system, although Asian medicine has never considered it anything but that. Everything, from the basic heartbeat, to the more complex micropulsations through which smell triggers an emotional response, to bone regeneration, to the mind's interaction with the immune system, as well as the individual cell's ability to replicate, is increasingly understood to have an initiating electrical component and complex interconnecting pathways.

We are very slowly—some would say too slowly—beginning to undergo a paradigm shift in Western medical thinking from a chemical-mechanistic model of the human body to a more finely tuned electrical-system model. Such a shift will revolutionize our notions of disease states and treatments. In the new framework, ancient healing systems like acupuncture are no longer lightly dismissed by mainstream doctors, more and more of whom offer it in their practices. Also, the use of magnets in treating a wide variety of ailments resistant to other therapies is currently being investigated at a handful of mainstream medical centers, often bringing results—especially for arthritis sufferers—that astound researchers. Such results are not supposed to occur, according to what we *thought* we knew about human anatomy.

Yet anything with the ability to heal also has the ability to harm, especially when its underlying biodynamics are imperfectly understood. The human body, when viewed as a sensitive electromagnetic organism with the potential to receive, transmit, and conduct minute signals, becomes capable of both healthy responses and pathological overload—much like a radio station affected by too much static in the atmosphere.

New so-called safe technologies like diagnostic magnetic resonance imaging (MRI) may be in the field prematurely. Short-term memory damage and other disturbing reactions are beginning to be reported. Approximately 20 percent of those receiving MRIs, for instance, experience severe panic, which has usually been attributed to claustrophobia. But claustrophobia does not afflict 20 percent of the general population; this wastebasket explanation does not suffice for the many patients with no previous claustrophobic episodes who panic inside MRI chambers. More likely explanations are that some people are extremely sensitive to electromagnetic fields or that the human body reacts badly to such fields. Questions of safety should play a more prominent part in risk/benefit considerations regarding all such high-tech equipment.

A Typical Scenario: Polarized Politics

Until the recent media attention, many laypeople became educated about electromagnetic fields through frustrating experiences that left them feeling powerless. Perhaps you, like thousands of others, have had a similar experience. . . .

You have found out that a radio or cellular-phone tower, an electrical substation, or a new high-tension line is planned for your neighborhood, and you recall hearing something about adverse health effects. Or perhaps you have read a little about electromagnetic fields and have started to wonder about the various satellite dishes and antennas near your house or apartment building. Or perhaps you live or work near high-tension lines or are in an occupationally exposed job, such as computer data work or law enforcement involving the use of a radar gun. Worse, you or a neighbor, a friend, or a loved one has been diagnosed with cancer, and you have tried to gather information about the possible cause.

But the more you read, the more alarmed you become. The more alarmed, the more you want to protect yourself—and then a puzzling thing happens. You come up against a wall of denial, resistance, and paralysis from clearly intelligent, well-meaning people in local, state, or federal agencies who you had assumed were there to help you.

To add to your confusion, for everything you read about adverse effects there is a well-credentialed scientist saying the exact opposite. It is as if you have walked into a hall of mirrors.

This is the world of EMF politics: an arena where ordinary citizens have to teach themselves physics and speak in odd terms like "centimeters squared" and "specific absorption rates"; where scientists are at each other's throats in a most unprofessional fashion, dismissing credible colleagues' research with a vindictive scorn rarely met with elsewhere in the sciences; where the military-industrial complex plays a shadowy yet monolithic role in the background; where Congress appropriates money for EMF research at leading federal agencies, which then refuse to spend it; where manufacturers turn out EMF-emitting devices in a regulatory vacuum, making no attempt to shield such products until lawsuits appear imminent; where mainstream medical practitioners know little if anything about EMFs and cannot answer your questions; where professionals from many different fields speak in confusing and patronizing tones, grossly underestimating the average person's ability to understand; and where just about everyone wants to play a surreal game of risk/benefit analysis with statistics that happen to represent your life. The very subject of electromagnetic fields embodies its own polarization metaphor: the positive and the negative aspects seem to cancel each other out, leading to a neutral state in which people are unable to decide what to do.

Any reasonably well-informed person must wonder what is going on here. Who is right? Surely some government agency is responsible for protecting consumers against EMFs. Where is it?

Those who actively engage the issue by trying to stop an EMF project or by seeking legislation to protect their community come up against an unsuspected layer of defense strategies. Community after community has to "reinvent the wheel" each time it tries to challenge the companies that profit from EMF processes or devices. In the end, such challenges leave many more unanswered questions and cynical citizens in their wake than there were at the outset.

But the questions that typically arise during these community debates are legitimate. How can homeowners protect themselves from drastic losses in the value of property located near high-tension lines? Why do law firms representing utility companies insist on gag orders as a condition for settling claims? Why must private citizens have to raise huge sums of money for legal fees to fight a project in one county, only to have the company pack up in mid battle and move the project to another county a mere five miles away, forcing its citizens to start all over again? Why must individuals sue manufacturers, only to have them fall back on government agencies, which refuse to set safety standards, allowing everyone to stonewall? Clearly, the situation is deplorable. The question is, how can we change it?

The Dynamics of Denial

When we think of denial what comes to mind is someone knowing the "truth" but pretending otherwise. But in fact, the process of denial is a complex psychological dynamic in which a person is actually not experiencing his or her own experience.

Denial has a useful, healthy purpose for the human ego

structure. When a loved one dies, for instance, it is a way to keep overwhelming fear and grief at bay until one can come to terms with the loss. A less healthy example of denial is the alcoholic who rationalizes continued drinking in various delusional ways.

Those who study the process of lying have observed that whole communities, organizations, and various official bodies can manifest group denial, which seems to occur most often when the subject under discussion is complex or has important implications for the status quo. Organizational denial is closely linked to another unproductive process that leads to group paralysis. Called "groupthink," this psychological dynamic can develop within a decision-making organization in which loyalty to the group or the need to reach a consensus becomes more important than solving the problem at hand. It can be a subtle phenomenon, which participating individuals often fail to recognize, but its by-product is the suppression, elimination, dismissal, or reinterpretation of dissenting information. Groupthink requires several forms of lying to maintain the status quo, including the ignoring or omission of facts, selective memory, and denial. The subject of electromagnetic fields has collided with organizational denial and groupthink more often than not, and continues to do so.

One would hope that any errors made were in favor of consumer safety, health, and property values, but this has not been the case, especially in the overtly probusiness practices of the Bush and Reagan administrations. The benefit of the doubt has automatically been given to the product, the manufacturer, or the public utility, with the burden of proving harm transferred onto the public—an approach that rests on the smugly cynical assumption that no layperson will be able to prove anything in the face of contradictory scientific evidence. (One way to maintain the status quo is to pit opposing scientists against each other.)

Fortunately, some of the institutional denial is beginning to dissipate. More and more EMF research is finding clearer answers; some governments around the world are moving to protect their populations (even as the U.S. government lags behind);

there are less-intimidating ways to talk about EMFs; businesses are taking mitigation designs seriously; the scientific climate is becoming less hostile, as more researchers are attracted to the subject and are refusing to be bullied; and, in general, a more educated public is insisting that people come before profits.

Unfortunately, change takes a long time. What can you do in the meantime to protect yourself and your family from unwanted exposure to the myriad EMF/RF-emitting devices that surround us all? How can communities protect themselves without turning back the clock to a pre-twentieth-century world? What is it reasonable to ask of those who profit from the communications towers and utility lines already in our midst?

This book will try to answer some of these questions.

The Military: How Did We Get Here?

The politics of safety concerns about electromagnetic fields follows strange corridors that twist through many competing interests vying for influence and power.

Conspiracy theories abound—theories that are interesting and bring an entertaining intellectual righteousness along with them. But such conspiracies are impossible to prove and ultimately are beside the point. To cause a gap in safety, all that a given situation needs is a vested interest protecting its own territory, without anyone consistently representing the best interests of the public, and the effect will be the same. What a consumer might see as a conspiracy and subsequent cover-up, a company executive might see as a good business decision or as damage control. Oftentimes the public interest just gets lost as societal values change. There are not always convenient villains to point to. The issue is not that tidy. It would be easier if it was.

That the military has a huge stake in EMF/RF research cannot be denied. The military gathers information through wireless technologies; creates weapons that project energy fields (although they currently deny they have developed such weapons for use against humans); has an array of infrared detection devices available; records, assimilates, and stores vast amounts of information; tracks aircraft and land-based vehicles on radar; makes infinitesimal calibrations; communicates over global distances — the list could go on and on. Most of the high technology around us was initiated by military research and development, and later adapted to civilian use. The U.S. military is the greatest source of both basic and applied science in the world, whether or not we agree with its approaches or with its end uses.

EMF research has been, and continues to be, conducted at various universities, in government laboratories, and by private industry. The door is a revolving one. Whole companies sprang up in the 1980s whose sole purpose was to take recently declassified raw military research and create civilian uses for it. Such companies then sold their product ideas to industry, whose own engineering departments developed the new product lines. The entire microchip and personal-computer age that has been ushered in over the last fifteen years came through this channel. So have cordless and cellular phones, as well as the "wireless" America that the major communications companies envision, which is already taking shape as cellular towers spring up like mushrooms in our neighborhoods. The national data base called Internet (the so-called information superhighway), which will eventually link on-line services all over the world, was developed by the Pentagon back in the 1960s. Pentagon research is also responsible for our ability to communicate via satellite within seconds from anywhere on the globe, making it possible for us to see the Gulf War and the Olympics live, for example. And Pentagon research is responsible for some low-tech items, too, like the popular lettuce spinner, a side benefit of the space program's studies of centrifugal force.

The military wields an enormous economic influence through the billions of dollars it awards to corporations and research subcontractors. In fact, some would say that the military and its subcontractors have been driving the U.S. economy since World War I. But driving the economy is not the same as protecting the individual consumer's health and safety, and the failure to make that distinction is what has gotten us where we are today.

Moreover, most military research is done by physicists and engineers, whose role is to figure out how to make things work, not by medical experts, whose province is to understand the causes and treatment of illness. The reintroduction of medicine into the area of EMF research is relatively recent. Medicine's absence was partly due to scorn from within the medical profession itself, which has consistently seen pioneers in EMF research become marginalized according to prevailing political whim, and partly due to a near absence of funding from outside the profession. That, too, is changing. Doctors continue to face a barrage of questions from their patients, which keeps their attention on the subject. Also, molecular biology, and such new interdisciplinary subspecialties as biophysics and bioelectromagnetics, are promising areas for meaningful research.

More epidemiologists, whose concerns include patterns of disease in population groups, have entered the field, too. And cellular research in laboratories is beginning to pinpoint biological tendencies but such research will not necessarily reveal the larger picture. Both the laboratory and the epidemiological research perspectives need to be utilized.

The Business and Legal Climate

Vast business interests in electromagnetic fields are also at stake—billions of dollars in wireless communications alone. In the short term, business stands to lose the most from stricter EMF standards and from government regulations requiring that devices be shielded, modified, or scrapped altogether for safety reasons. But the long-range perspective is a profitable one, especially for companies that offer safer products first.

It is heartening to see large professional conferences specifically addressing bioeffects and mitigation techniques for consumer products and for high-tension lines. There was even a conference on magnetic resonance imaging held in 1991. And most computers manufactured today are lower in emissions than the older machines. In the absence of U.S. standards, most major computer manufacturers are adhering to the strict Swedish standards.

But businesses are caught in what they perceive as an impossible legal and public-relations bind. They cannot develop and tout low-radiation models (which for many products, including computers, entails nothing but inexpensive shielding materials) without in effect admitting that their existing products might be harmful. Some employers resist providing shielding devices or special clothing for EMF-exposed employees, lest they imply that working conditions are unsafe. Many captains of commerce prefer to hide behind the current contradictions in scientific findings. Others are acting more responsibly and are developing new products and strategies.

And a longer-range business calculation is emerging. The reality is that manufacturers, utilities, communications companies, and other employers would like to be rid of the high attorneys' fees for litigating EMF cases. These damage suits are considered the "toxic torts" of the 1990s, and their number will only increase as time goes on. The business world is well aware of the pitfalls, as hinted at in one New York trial judge's recent ruling that the normal statute of limitations does not apply to those filing claims for EMF health damages (this has long been true of claims for asbestos and other toxic substances).

Gag orders, which seal the record of a case or a settlement from public scrutiny, are a common attempt to limit access to the growing data base, making it more difficult for plaintiffs' attorneys to readily tap into the available information, especially with respect to real estate and health claims. Although less often used for siting cases, gag orders can also prevent information from being passed on by one community to another when a cellular tower, utility project, or substation has been successfully challenged. As consumer and legal awareness grows, such tactics will be less successful.

One less-obvious component of the EMF health issue is easy to lose sight of in our haste for sensational exposés. Many people of good conscience make their livings in EMF/RF-related companies and would not willfully or knowingly harm others. It is insulting and inaccurate to presume that everyone working in the utilities industry, for instance, is unethically hiding information from the public. Many of these workers would like answers themselves, to protect them and their own families. Some of the best sources of information have, in fact, been people in EMF-related industries. It is also a common practice among electric-utility companies to send inquiring consumers lots of information, both pro and con, to let them make up their own minds (though the companies can sometimes take a highly belligerent stance).

Unfortunately, the telecommunications industry is ten years behind even this pro-and-con approach. These companies typically use stonewall tactics or hide behind their Federal Communications Commission licenses as a shield against consumer-safety complaints. This, too, will change. A handful of European insurance companies, perhaps recognizing that physical problems are associated with some EMF frequencies, are said to be charging higher rates to users of cellular phones. American insurance companies might not be far behind.

Chapter 2

THE DIFFICULTY WITH
SETTING STANDARDS

EVERYONE—consumers, scientists, government agencies, and manufacturers alike—would like to have viable, useful EMF safety standards that are based on solid, reproducible science, so that we all could relax. It is a simple, reasonable request, but tough to accomplish, for several reasons.

Scientific data in any field is always open to various interpretations. Sometimes two researchers can take similar test results and arrive at radically different conclusions, both of them logical. Plus such data rarely produce a clear demarcation line between a hazardous level and a safe level. It is often difficult to distinguish a physical reaction within normal ranges from a hazardous one. So where should a protective standard be set? At the point where any reaction is observed? Or at the point where a harmful reaction might be anticipated?

In addition, standards rarely deliver the margin of safety that they seem, by their very existence, to imply, because different people respond very differently to the same physical stimulus. What's deleterious for one may have no impact on another. Cigarette smoking is a good illustration, and comparable to the EMF issue in several ways. We generally accept that smoking causes various life-threatening cancers and illnesses, yet some smokers live to a ripe old age. As with EMF, no one knows the precise physical dynamic or trigger that initiates cancer in smokers. But the medical and scientific community has nevertheless reached a consensus that smoking is harmful. And, as with EMF industries, the powerful tobacco lobby has overly influenced governmental policy for its own benefit and profit.

Although no one knows yet what type of EMFs or what specific properties of them are biologically dangerous, several physical factors come into play regarding the bioeffects of various frequency exposures: individual genetics, duration of exposure, distance from the exposure source, presence of other possible toxins, an individual's size, weight, sex, and age. A child's anatomy reacts quite differently than an adult's, and teenagers' hormonal changes affect their reactions.

Despite these complex variables, it *is* possible to set standards, and it *is* reasonable for citizens to insist that their safety be of primary concern both to their government and to the companies that enjoy the privilege of doing business with them.

A Case in Point: The History of ANSI C95.1

The safety standard most widely used in the United States since 1966 is the American National Standards Institute (ANSI) C95.1 guidelines for microwave exposure. The history of the adoption of these recommendations illustrates the gamut of troubles in establishing such standards.

Many agree that the guidelines are woefully inadequate, both in kind and in degree, and that merely to tighten them is not enough, given the scientific picture that is emerging. In fact, some of those who originally helped create the standard had doubts at the time that have yet to be properly addressed.

It is important to note that the ANSI standard is for the *microwave* frequencies used in communications systems and radar, among other things, and does not apply to the lower-end frequencies such as those common to electrical use.

The C95.1 standard had its origins in the early 1940s, stemming from the fears of military personnel about the safety of radar. It was widely known among U.S. servicemen that radar could produce temporary sterility (some even used it as a form of birth control). In 1942, the U.S. Navy's Bureau of Ships directed the Naval Research Laboratory to investigate the possible harmful effects of microwave radiation. Within a short time, other military agencies, including the National Defense Research Council, the Aero Medical Laboratory of the Air Technical Service Command, and the Army Air Field at Boca Raton, Florida, became involved. All found that there was no reason for alarm but that caution should be used to avoid prolonged exposure. No specific guidelines were recommended.

This situation must be understood within the context and social values of the time. World War II was a popular war, and radar was widely regarded as beneficial to the national effort. Concerns for individual safety were sublimated to the daunting task at hand. It was also several decades before consumer protectionism became a nationwide movement.

Immediately following the war, microwave equipment that had been developed during it, by companies like Raytheon, became available for medical research, which was then looking into the deep heating of selective body tissues through diathermy. Interest shifted from the study of potential hazards to therapeutic applications, and the need for any safety standards was dropped.

Again, in the context of the times, people deferred to scientists in general, and to medical experts in particular. Medical opinion was that, if used judiciously, microwaves were a safe, convenient, and comfortable modality for localized deep-tissue heating. This opinion prevails today, despite the fact that it is based on a thermal-effects-only model that is most likely inaccurate and is based on very limited research.

Although very little research on the effects of microwaves was conducted after the war, in 1948 researchers at the Mayo Clinic made the first confirmation of adverse effects solely attributable to microwave exposure, the formation of cataracts in dogs. Simultaneously, researchers at the University of Iowa, at the request of Collins Radio in Cedar Rapids, Iowa, a subcontractor for the militarily sponsored Rand Corporation, also reported a possible link between microwaves, cataracts, and testicular degeneration in dogs. These findings attracted little interest, however—especially from funding agencies.

Thus began a disturbing pattern of companies taking an interest in basic research, only to withdraw their support if the results proved disfavorable to their needs. It has been fertile ground for conspiracy theories. And it has been a very real source of professional anguish for some of the country's most farsighted researchers, who have regularly had research monies granted and then withdrawn. It has also been a linchpin for maintaining the status quo.

Interest in the bioeffects of microwaves was rekindled in 1953, however, when an alarming report was issued by John T. McLaughlin, then a medical consultant to the Hughes Aircraft Corporation. Concern had persisted over adverse health effects on radar workers. McLaughlin's report, sent to the military, listed for the first time internal bleeding, leukemia, cataracts, headaches, brain tumors, heart conditions, and liver involvement with jaundice as possible effects. (McLaughlin also later consulted on the first suspected death from radar exposure.) The military response was immediate, with the Air Research and Development Command directing the Cambridge Research Center to investigate the biological aspects of microwaves with an eye toward determining tolerance levels for both single and repeated dosages. The Navy also convened meetings on the subject.

But little empirical data were available on which to base tolerance levels, so the Navy group tried another route: establishing the amount of radiant energy that the human body could absorb and then eliminate through normal body functions. In a complex approximation that took the normal size of the body to be that of an average male weighing 70 kilograms and having a surface area of 3,000 square centimeters, it was estimated that about one-third of the radiation from any source could be absorbed and eliminated through normal body functions, meaning that the body could cool itself down if it became artificially heated. A safety factor of 10 was built into the calculation, and the baseline figure was set at an exposure of 0.1 watt per square centimeter, or 0.1 W/cm². (This is a unit of measurement representing power deposited over a set amount of space.)

But it was soon apparent that the calculations were way off. Herman Schwan, a biophysicist at the University of Pennsylvania's Moore School of Engineering, sent a memorandum to the Office of Naval Research pointing out that the amount of heat the body dissipates under normal circumstances is 100 watts, not the 150 watts of the original calculation, and that the absorbing surface of the body is closer to 20,000 cm² rather than 3,000 cm². The net effect of absorbed radiation, Schwan figured, was more than twenty times greater than the body could dissipate. The standard was soon lowered to 10 milliwatts per square centimeter (10 mW/cm²), and this became the basis for the C95.1 industry recommendations of 1966.

During this time, industry was also trying to set guidelines for their employees. Bell Telephone Laboratories and General Electric, both huge military contractors, sponsored meetings that put more emphasis on empirical data than the military had. Particular attention was paid to the 1952 work of Frederic Hirsch of the Sandia Corporation in describing cataracts, or lenticular opacities, that developed in laboratory technicians regularly exposed to microwaves at power levels of around 0.1 W/cm², the exposure at which actual damage had been observed.

In 1954, General Electric set a stricter standard by a factor of 100, at 1 mW/cm². Bell Telephone developed a standard 1,000 times greater. In essence, both industry and the military agreed that physical damage (in the form of cataracts) could occur at 1 W/cm². Where they differed was over how large a safety margin should be built in. But hazards were thought to be due solely to the ability of microwaves to heat tissue, called thermal effects. Other biological effects were largely ignored, and the subject continues to be of scientific debate today.

These early safety margins were never intended to be the final word, but only the beginning of the search for long-term solutions for microwave exposures. They were intended to provide conservative estimates until enough data had accumulated to confirm or deny the estimates. Over and again, more research has been called for. This is still true today, fifty years later.

During the 1950s, the major sponsor of research became the U.S. Air Force, which carried out this work in conjunction with the development of the newest microwave technologies. Although there is a certain rationale for having everything under one agency roof, with the military developer of the technology also being the first to see its physical ramifications, some would say that it was more a case of the fox guarding the chicken coop. This became especially true in the 1950s cold-war atmosphere of secrecy rather than disclosure.

Unfortunately, not much of that military perspective has changed over the years. The Air Force appears to dominate the research field today, and where its technology intersects with the general population there continue to be issues of grave concern. In 1991, for instance, residents of Taos, New Mexico, began to hear a strange vibrational hum that caused some to suffer sleeplessness and headaches. Behind the noise were a structure suspected of being used to bounce radar waves and a low-flying airborne laser laboratory. The military has denied any possible connection. All over the country, civilians who are worried about radar installations find it all but impossible to get information from the military, and military radar operators who develop health problems after leaving the service are uniformly stonewalled.

By 1956, a massive four-year research effort called the Tri-Service program had been initiated, under Colonel George M. Knauf. The objectives were impressive: the program was to clear up any unknowns about microwave exposure, with an eye toward understanding the mechanisms of microwave-tissue interactions. The aim was a broad-based search to determine the range of biological effects, combined with an effort to assemble all pertinent empirical data in order to set safe-exposure standards. The research was conducted in Rome, New York, at the Air Research and Development Command (ARDC). Unfortunately, the Tri-Service program never addressed this last, standards-setting goal in any formal way, despite claims to the contrary, and the information it gathered remained largely basic, rather than applied. The issue of standards setting passed to the Navy, and to the few segments of industry that remained interested.

In 1957, the Department of Defense ordered the U.S. Navy's Bureau of Ships to conduct testing for microwave hazards, and in 1958 the bureau was given the responsibility for setting standards. It conducted several categories of inquiry, one of which, pertaining to personnel, was reassigned to the ARDC in Rome. This brought the Tri-Service program loosely under the Navy, which also turned to the industry-supported American Standards Association (ASA), a precursor of the American National Standards Institute (ANSI). A special committee called C95 was formed within ASA, and its work was jointly sponsored by the Bureau of Ships and the American Institute of Electrical Engineers (AIEE), now called the Institute of Electrical and Electronics Engineers (IEEE). C95's first meeting was in February of 1960.

The committee's work was slow and contentious, and in retrospect it has all the hallmarks of that unproductive process called groupthink. Herman Schwan was finally chosen as chairman, six months after the committee's formation, and he was only reluctantly accepted by the AIEE. Several subcommittees were set up within C95 itself, and it took several years to produce the first draft of recommendations, called C95.1, which came from one of the subcommittees that Schwan had taken over. The draft was then circulated to all members for a vote, but not until November 1966 was the draft officially adopted. The delay was apparently due to the failure of some voting members to return their ballots in enough numbers to reach a quorum on time—the number required to reach a quorum was eventually lowered instead. By the time the recommendations were adopted, Schwan had resigned his chairmanship. The recommendations, called ANSI C95.1-1966, set the safety standard at 10 mW/cm².

Problems with the standard were obvious almost immediately. The military had problems of implementation. The standard was set with high-intensity occupational exposure in mind, not the general public, which could experience low-level, long-term continuous exposure. Some in C95 had called for a separate standard for the public, which was never developed. And doubts existed about the research on which the standard was based.

This research had focused on clinical studies and personnel surveys, on animal experiments, and on studies of anomalistic effects, or nonthermal-based reactions. The personnel studies promised the most valuable information and were widely touted to have the most definitive findings. But closer examination showed that these early 1940s studies, which found no evidence of abnormalities, had been given far more credence than they deserved. The studies were extremely superficial, gathered no urinalysis or blood-chemistry data, and so their conclusions of "no harm" could not realistically have been reached with the data at hand—especially when compared with other studies that found a significant concentration of immature red blood cells, as well as a high incidence of headaches, in exposed workers. A case could surely have been made for better test protocols, and over the years some tried to prod the military into improving its procedures. But neither the quality nor the quantity of the personnel studies changed much, although the topic kept coming up. Industry, meanwhile, was willing to cooperate but was waiting for someone else to take the lead.

A few contradictory studies were carried out during the 1950s. Researchers at Lockheed Aircraft found blood abnormalities in some personnel but later dismissed the findings as a statistical variation. Milton Zaret, a clinical professor at New York

University-Bellevue Medical Center, however, found a disturbing pattern of cataract formation in radar workers, in deep areas of the eye not normally prone to cataracts. He reported finding posterior polar defects, luminescence, and early opacification in workers far too young to have such disorders. Zaret thought that thermal effects were not the only mechanism for what he observed and that low-level exposures were cumulative. More recent research in Croatia also found posterior cataracts in exposed workers, verifying Zaret's work.

Animal studies, though better controlled and more extensive during that time, neither supported nor refuted the 10 mW/cm^2 guideline. Numerous studies found no irreversible damage in animals exposed to radiation in excess of 10 mW/cm^2. Most of this research had been farmed out by the military to universities — the State University of New York at Buffalo, the University of Rochester, the University of California at Berkeley, the University of Washington, and the University of Miami, among others. (Some of the same university researchers were simultaneously consulting for industries developing civilian technologies like microwave ovens.)

Without clear data, the animal studies were expanded upon and used to support three conclusions: any biological effects were thermal, noncumulative, and of little concern. That is still the prevailing view behind today's technology, encompassing everything from cellular phones and high-tension lines to ultrasound and MRIs. But this is changing.

Although widely embraced at the time, consensus was certainly not unanimous within the scientific community during the Tri-Service era. The experimental techniques of the animal studies were called into question. Findings were often not replicated, and, although not as numerous as the negative reports, some anomalistic findings were reported that suggested nonthermal reactions. In addition, dose levels often went unreported in the studies, and when they were, the source output was measured but not the intensity of the field or the specific tissue absorption.

Nor were allowances made for such important factors as continuous wave (a steady exposure) versus pulsed (intermittent) irradiation. Also, the studies themselves were conducted at high power for short durations, the exact opposite of what occurs among the general population, whose exposures are long term at low power.

In other words, some research work had been done, but it was of a kind inappropriate for the conclusions it was used to reach. Many now say that an accurate understanding of microwave's bioeffects should never have been inferred from this data, much less have been considered complete enough to allow technologies to proliferate within these frequency bands. Study projects of low-level, long-term effects have been repeatedly called for over the years, but only one was initiated and completed.

In the early 1980s, the U.S. Air Force School of Aerospace Medicine funded a large and costly study at the University of Washington, under the direction of Dr. Arthur W. Guy. Rats continuously exposed to high-frequency microwaves, which were nevertheless twenty times lower than what was considered the "safe thermal level," were found to have more primary malignant tumors (eighteen exposed animals) than nonexposed control animals (five). This study itself became controversial; it will be discussed in more detail later.

Even in the early years, military researchers had seen disturbing nonthermal effects, including testicular damage in rats at exposure levels near the C95.1 guideline, abnormal brain responses, and changes in blood count. Some early 1920s research had found increased growth rates and other deleterious effects on malignant tumors and certain bacterial toxins.

Many involved with C95.1 objected to setting the recommendation at that time, given the state of the scientific uncertainty, but it was adopted nevertheless. Some say the adoption of the standard was more to please the military than to address genuine safety issues. Unfortunately, once a standard is in place, it is likely to become entrenched.

There had also been differences between the defense industry and the medical community. At any given time, three research perspectives existed: clinical, biological, and engineering/physics. Engineering/physics won out over the other competing interests, none of which represented the general public. Some companies, like Hughes Aircraft and Lockheed, were looking for serious answers; others were rumored to be pressuring the military to withdraw funding for cataract research. A physician-researcher like Milton Zaret could report serious indications of damage, while another clinician offering diathermy treatments could say something entirely different. (This is partly due to the inherent difference between clinical practice in general and basic research in the laboratory.) Amid all these competing and conflicting interests, the public continued to be left out of the picture. Microwaves were seen as synonymous with radar, which was the province of the military and its industrial contractors. That consumer products emitting microwaves were rapidly being developed by these same companies somehow got lost in the shuffle.

There is little doubt that possible evidence against microwave safety was ignored and that research which could have shed some light was not undertaken. But in the context of the times, the larger objectives of national security, rather than individual welfare, strongly influenced the decisions. Many decisions were presumably made in good faith (although military intentions during that period are difficult to assess, especially in light of recent revelations about the use of unsuspecting civilians in nuclear-radiation research). But it is not good faith forty years later, when we have less military pressure. Nor does it help to know that a relative few like-minded people controlled most of the early decisions about the direction of research, which undoubtedly limited the input from other, equally valid perspectives. These need to be factored in today, in a renewed effort to address the public welfare.

Standards: Where We Are Today

It is important to note at the outset that ANSI (American National Standards Institute) is an industry-controlled organization composed of several committees that set standards for many products, devices, and processes, not just for radiowave/microwave transmissions. (People knowledgeable about the microwave controversy refer to "the ANSI standard" as if there's only one standard; this is misleading.) Compliance with ANSI recommendations is usually voluntary, although some segments of industry mandate them.

The IEEE (International Electricians and Electronics Engineers) has reviewed its basic ANSI C95.1 recommendation periodically. In 1982, it sent recommendations for radiofrequency exposures to ANSI that lowered the 1966 limit of 10 mW/cm^2 to 1 mW/cm^2, but only at certain frequencies. Today it remains an extremely complex standard. Having been rewritten again, it is now called ANSI/IEEE C95.1-1992.

The revised versions of the standard, unlike past versions, set a two-tiered recommendation, one for the general public and one for workers in the radio-frequency field, thereby acknowledging a difference in absorption rates. But the model used is still the "adult of average weight and height," and the standard still presumes only thermal effects. Although the two-tiered recommendation is a step in the right direction, many say that the thermal-effects-only standard is wholly inadequate and that the same old problems still exist. Moreover, in a new revision now being considered by the Federal Communications Commission, some high-frequency exposures are allowed a twofold increase over the previous version of 1982—clearly a step in the wrong direction.

But individual states, industries, and others have set standards more strict than the ANSI/IEEE guidelines. The increasingly lower limits recommended by ANSI over the years are hardly radical proposals. Some states have set lower population-exposure standards for years—in other words, improvements are do-able. In addition, both a U.S. Air Force research group at Kirkland Air Force Base in New Mexico and a group at Hughes Aircraft Company have recently adopted limits for human exposure that are a hundred times more strict than the revised ANSI recommendations. The reason given for the need to tighten the standards was an abundance of nonthermal effects that had been recorded in the scientific literature worldwide by the end of the 1980s.

Various professional organizations and governmental agencies have attempted to set radiation standards for different frequencies, some more lenient than others. The International Non-Ionizing Radiation Committee (INIRC) of the International Radiation Protection Association (IRPA) has been developing health criteria for nonionizing radiation, in cooperation with the Environmental Health Division of the World Health Organization (WHO). INIRC has issued a series of "Environmental Health Criteria" for ultraviolet radiation, radio frequency/microwaves (RF/MW), ultrasound, lasers and optical radiation, extremely low frequency fields, and magnetic fields. Compliance with any of its recommendations is voluntary; many of its committees are made up of differing ratios between industry and academia; many of the committees presume thermal effects only.

A recent WHO report titled *Electromagnetic Fields (300 Hz – 300 GHz)*—the RF/MW band—reaffirms guidelines set by IRPA in 1984 and 1988. (IRPA assembled the committee that wrote the WHO report.) Committee members come from a wide range of countries with very different perspectives and sometimes conflicting interests on the subject. IRPA is now called the International Commission on Non-Ionizing Radiation Protection (ICNIRP.) Those wishing to get a copy of the WHO report should contact WHO Publications Center, 49 Sheridan Ave., Albany, N.Y. 12210, and ask for order number 1160137. The report is technical and not for the average layperson, but if your local municipality is looking for information upon which to base standards, they should have the resources to hire technical support.

Unfortunately, this report, too, presumes thermal effects only, but it at least raises the possibility of carcinogenic effects and calls for more studies. Some critics say it does not go far enough in protecting civilian populations.

The American Conference of Governmental Industrial Hygienists has, for the first time, published threshold limits for industrial exposure to magnetic fields—another step in the right direction. But its recommendations are more lenient than most in that they allow greater exposures.

The number of official-sounding organizations with mind-deadening acronyms is numbing to consumers and lawmakers alike who just want the bottom line: which standard is the most protective? Critics say all are flawed in one way or another, but the best guidelines at present appear to be those recommended by the National Council on Radiation Protection and Measurements (NCRP) in its Report No. 86, titled *Biological Effects and Exposure Criteria for Radiofrequency Electromagnetic Fields*. These set exposures for the general population that are five times more strict than the ANSI standard. (See Chapter 15 for more information on RF standards.)

The NCRP is the only organization chartered by the U.S. Congress to develop radiation protection recommendations. It publishes a number of commentaries and reports for fees ranging between $10 and $60. The phone number is 800-229-2652 for more information. (You will need to know specific report numbers before ordering, but NCRP's publications department will send you a list.) NCRP's interests encompass a broad range of frequencies, and it has set up umbrella committees to examine nonionizing radiation used in medical lasers, MRI and vivo spectroscopy, hyperthermia procedures, and ultraviolet light exposures, among others. Their scope includes measurements, exposure assessment, biological effects, mechanisms of interactions, public and occupational exposures, and mitigation techniques. Their standards for RF/MW exposures are considered the most protective available, and communities can adopt them without FCC approval at the present time.

There is some question about whether federal agencies have the right to pre-empt state or local zoning with regard to public exposures. The FCC has thus far refused to insist that its standards prevail over local standards. But it is a two-edged sword. A community can set standards as strict as it likes, but that results in a patchwork of regulations across the country. This is fundamentally ineffective, since lenient standards in one community will adversely affect neighboring communities. Good, effective national standards are really what is needed. Most communications companies would also prefer national standards rather than having to deal with local regulations. Unfortunately, the communications companies would prefer *lax* national standards and have recently asked the FCC to override all local jurisdictions.

The Environmental Protection Agency, the Federal Communications Commission, and the Food and Drug Administration

Perhaps of most significance to consumers is what the Environmental Protection Agency (EPA) has experienced in the matter of standards. The EPA is the government agency that should have established guidelines for radiation exposure, in line with its responsibilities for protecting public health and the environment. The reasons it has not are typical of EMF politics.

At one time, the EPA had a well-respected laboratory that conducted its own research, but it was largely gutted in the 1980s, during the Reagan and Bush administrations. The agency became highly politicized by the conflict between business interests and those of environmentalists and citizen-welfare groups. Enterprising companies with designs on mineral, water, and timber rights, for example, came to see our natural resources as giant candy stores. Telecommunications companies saw the nation's airwaves the same way. Many dedicated people left the agency during this period.

The EPA had been promising for years to set RF/MW guidelines for the general public, but it completely abandoned the effort in 1988. It had tried for over a decade to develop regulations and had even circulated a draft proposal for preliminary review at one point. (In 1991, its own Science Advisory Board recommended that it finish the guidelines it had started. It intends to do so soon, but the guidelines will not have the force of law.)

Critics say the EPA has been seriously derelict on this issue, but spokespeople within the agency say it just got tired of the political shenanigans and wanted to co-ordinate its efforts with the White House before proceeding further. After significant pressure from a number of quarters in recent years, however— including a request from President Clinton that the agency issue a report on possible cancer links with EMFs—the EPA recently decided to reactivate a program on nonionizing electromagnetic radiation that had been almost completely abandoned by 1993. Work will again focus on EMFs and RF/MW radiation, let's hope with a more comprehensive strategy than previous efforts displayed.

The EPA does not deal with the extremely low frequencies (ELF), such as those around power lines, but rather with the higher radio and microwave frequencies that are around communications towers. Nevertheless, in 1990, the agency published a preliminary draft report called *Evaluation of the Potential Carcinogenicity of Electromagnetic Fields* in which it recommended that low-frequency electromagnetic fields be considered a class B carcinogen, like radon. The review was carried out by the Office of Health and Environmental Assessment within the Office of Research and Development, at the request of the EPA's Office of Radiation Programs. The purpose was to explore risk factors for cancer from exposure to nonionizing radiation. While the entire electromagnetic spectrum was of interest, the report focused on two ranges, one encompassing everything from domestic wiring to high-tension lines and the other including most telecommunications-type exposures. When it appeared that the Bush administration was going to quash the report, the draft review was released, and it became one of the most quoted "preliminary" reports around. The EPA's Science Advisory Board intends to release a shorter, revised version, which is said to call for an emphasis on nervous-system and biophysical effects. In the meantime, however, the extremely low frequencies have been taken out of the class B carcinogen category. It remains to be seen how or if this will pertain to RF/MW frequencies.

On the RF/MW front, the Federal Communications Commission (FCC), which licenses all radio and telecommunications activities in the United States, has been using the ANSI/IEEE C95.1-1982 recommendations in connection with its licensing provisions for several years. The FCC is now considering the adoption of the 1992 version, as well as other standards proposed by NCRP and IRPA. Without EPA guidelines, the FCC has been caught in something of a vacuum. Although standards better than ANSI's have been available (such as the NCRP's), the FCC has consistently used the ANSI standard. To its credit, the FCC at least made adherence to the 1982 version mandatory rather than voluntary; to its discredit, the agency does not regularly monitor licensees for compliance. Transmission exposures are computer estimates done by individual tower operators. The FCC will come out to monitor a site only if it receives complaints of radio or TV interference.

The EPA has called the revised ANSI/IEEE standard "seriously flawed" and has questioned its ability to protect the public's health and safety. The agency specifically cited the standard's failure to recognize nonthermal effects and its different limits for "controlled" and "uncontrolled" environments, which are not directly applicable to general exposure. It has also criticized the twofold increase in maximum permissible exposures at high frequencies over what was allowed by the 1982 version. The EPA has called on the FCC to adopt instead the 1986 NCRP exposure limits contained in NCRP Report 86, and it also recommended that the FCC request the NCRP to update its reports, which the NCRP is doing.

The Food and Drug Administration (FDA) is another government agency with a stake in standards for nonionizing radiation. Its Center for Devices and Radiological Health (CDRH) contrarily finds the new ANSI/IEEE-1992 standard "acceptable" for transmission towers but objects to an exclusion that would exempt from the RF standard lower-power items that come under FDA supervision. Such items include hand-held, portable two-way radios like walkie-talkies, cellular phones, and personal communications devices. The FDA wants more research on such devices and has criticized the cellular-phone associations' carte blanche statements about safety. The FDA notes that such devices can induce relatively high specific absorption rates in the part of the body closest to the phone. (It has issued an advisory warning to cellular-phone users to keep their conversations short.) On the surface, it looks as if the FCC is trying to keep the responsibility for such low-powered transmitters (which differ greatly from communications towers, with their high-power output) on the FDA, which does not want the entire regulatory burden. Many expect that the FCC will discontinue its exemption of such devices from standards and testing in the near future.

The issue of which standards the FCC adopts will be an important one over the next few years. Communications towers transmitting at various frequencies are proliferating at an alarming rate, often in close proximity to homes and schools. The FCC recently sold whole frequencies of the electromagnetic spectrum in the gigahertz range to several huge telecommunications conglomerates, for the development of wireless personal communications systems like computers, faxes, modems, and phones that do not require electrical connections. There was no national debate about the propriety of the FCC's selling off the national airwaves similar to the debates that occur when other natural resources (oil, timber, minerals) are sold.

It seems to be one step forward and two steps backward. There are no government regulations that require appropriate scientific data before a new course of technological action is undertaken that could have detrimental bioeffects. The FCC, by its own statement, is a licensing and engineering body that relies on other agencies to conduct research on bioeffects and set standards. Consequently, the FCC can grant a license, and a user can develop a territory, often close to a population area, and still no regulatory agency is truly protecting the public's health.

Many of those hoping for substantial progress in our understanding of nonionizing radiation's bioeffects recommend that all research be consolidated under one organizational roof, like the National Institutes of Health, since the matter is really a medical one. Others say it is up to the EPA. Everyone admits that there is chaos about which something needs to be done.

The U.S. Congress: Toward Clearer Perspectives

Congress has gotten ever more involved with the EMF/RF issue over the years, largely due to prodding from the voting public. Various bills have allocated research money to different agencies, and subcommittees have looked into specific subjects, such as a subcommittee investigation, headed by Senator Joseph Lieberman, a Connecticut Democrat, into the adverse bioeffects of police radar guns.

In 1992, Congress set aside a budget of $65 million for EMF reasearch over a five-year period. (Industry must come up with matching funds.) Called the EMF Research and Public Information Dissemination (RAPID) Program, it is jointly run by the Department of Energy (DOE) and the National Institute of Environmental Health Sciences (NIEHS). RAPID appropriations are first given to the DOE, which then allocates some to the NIEHS. (The DOE also conducts its own EMF research, but critics liken this to the fox guarding the chicken coop.) Bureaucratic infighting has caused delays but NIEHS recently awarded $15.5 million for 21 research grants for animal and cellular studies. Allegations of bias in evaluating research proposals have been made, and the peer-review panel that has been assembled by the

National Institutes of Health has also come under attack. In addition, with government cutbacks, the fate of the DOE itself is uncertain. RAPID research is concentrated in the 60-hertz and ELF ranges, some of which might be applicable to the RF/MW frequencies in time. But a similar government research progam for the RF/MW frequencies is now more imperative than ever, especially with the advent of digital technology and the proliferation of RF/MW consumer devices.

There are also proposals before Congress that call for all EMF devices to be labeled for emissions, the same as food products are labeled for nutritional content and paint for chemical solvents. It is a terrific idea that will put the purchasing and/or use decision in the hands of the consumer. (The legislation would affect only consumer products; it would do nothing about transmission towers.) There would undoubtedly be complications at the outset, but such labeling would eventually be in everyone's best interests, including the manufacturers. One of the legal avenues that attorneys are using in product-liability suits is a "failure to warn," and labeling would help avoid that accusation.

It is imperative for us as consumers to let our congressional representatives know that the subject of EMFs is important, that it won't go away. It is also necessary to let them know that it is a bipartisan issue cutting across party lines, that it would be helpful if federal agencies like the EPA were free to do their jobs without political interference, and that the FCC needs to understand that its social responsibility to the public goes beyond a mere licensing and engineering function.

It is widely felt that the matter of public exposure to electromagnetic fields is going to be one of the most important issues we face in the next ten years. Billions of dollars are at stake. The implications are profound, and will touch every aspect of twenty-first-century life.

Nontraditional scientific perspectives and humanistic values should be factored into both the research and the setting of standards. But the work of those who conduct the research and those who set the standards should not be so closely intertwined. These are separate and distinct activities, which our lawmakers often confuse.

The aim of research is to understand natural processes and the mechanisms of interactions and to gain detailed knowledge about anomalistic phenomena. It should not be mission-oriented toward military or industrial applications or have standard-setting prerogatives.

Standards setting, on the other hand, does not require a full understanding of underlying physical mechanisms, but only data showing at what level effects, harmful or otherwise, appear. To mix the two dilutes and misdirects the efforts of both.

Also, our lawmakers should not rely quite so heavily on the scientific community for fundamental direction. Science is concerned with observation and experiment, probabilities and certainties. It is not trained to engage in or anticipate debates over social values that may arise as a by-product of the work. Sometimes there is no one correct public response resulting from a scientific development. The advent of nuclear weapons is a good example; it was not the province of scientists to raise all the pertinent public-policy questions regarding their use. That kind of evaluation is for us and our lawmakers to do.

The role of the consumer, in the meantime, is to keep pressure on our lawmakers; to insist that industry's special interests be kept in perspective and at a minimum and that conflicts of interest be abated; and to make sure that consumer interests are not lost sight of again. Moreover, industry and the military must be held accountable for their actions, legally and otherwise.

So far, government has been content to let business recommend its own standards for EMF emissions and to make compliance voluntary. That is no longer satisfactory. Public pressure has become too great; consumer suspicions about conflicts of interest are well founded; and the only standard widely in use for communications frequencies, which affect us all—the ANSI/IEEE C95.1 standard—is now and always has been inadequate.

Chapter 3

ELECTROMAGNETICS MADE SIMPLE . . . IT'S A LOT LIKE BASEBALL

D O YOU EVER wonder how a remote-control device works? Or what occurs behind the wall when you flip a switch and the lights come on? How the TV picture gets from the nightly news commentator at the station onto your screen? How your car radio can play without being plugged in? Or, when a radio is plugged in, how the voices carry through space the way they do? What about satellite transmissions? And smoke detectors—do they have tiny mechanical noses inside? What about automatic door openers? Do they really "see" you coming? In fact, how do any number of the devices that we take for granted every day work? It's amazing that so few of us do wonder about them, which is proof of just how much we take for granted. In other times, all of today's technology would have been considered very powerful magic indeed.

To understand better how these devices work, and to understand some of the distinctions and recommendations that will be made later on, you need to know a few basic scientific principles. This need not have the dreaded effect that science courses may have had on you as a student if you remember that you are reading it by choice, for no other reason than to better understand the physical world around you.

Atoms: The Fundamentals of Energy

All of life, as well as the fundamental building blocks of non-living materials, is the manifestation of energy in motion. Even something as seemingly solid as a table, when observed at the molecular level, is a grouping of similar atoms in motion.

Atoms are the smallest units of all matter. Within the atom are smaller units still, called electrons and protons, which are equal in number to each other. What makes one element different from another is the number of electrons its atoms have in orbit around the center of the atom, which is called the nucleus. (Hydrogen has 1 electron, helium has 2, and so on through the elements, up to 103 electrons for lawrencium.)

The nucleus of an atom, which contains almost all of the atom's mass, is dense and positively charged. The nucleus has its own properties and consists of protons and neutrons. Electrons have a negative charge while protons have a positive charge, and neutrons are neutral. It is the balance of the positively charged protons and the negatively charged electrons that keeps them in dynamic orbit opposite each other within the atom. The atom normally remains undivided in chemical reactions, except for limited activity in the removal, transfer, or exchange of certain electrons. We once thought it was indivisible, but we now know that splitting the atom releases tremendous energy in the form of nuclear explosions.

Atoms can combine into larger aggregate groups, called molecules, by lending, trading, or sharing their electrons. Each chemical compound is made up of molecules of one specific kind.

Electrons can move up and down in different orbits within an atom or molecule when they are "excited." This releases energy in various forms. Visible light, for instance, can be emitted or can be absorbed when electrons are excited into lower or higher orbits, respectively. Electrons can also be shaken loose from atoms and, when allowed to travel through a metal wire, they produce ordinary electricity. Depending on what the electrons are up to, the various physical phenomena they produce can be classified as chemical, optical, or electrical, but the nucleus itself remains relatively inert.

Besides electrons, neutrons, and protons, atoms are made up of other, very complex entities that we call subatomic particles. Electrons, for instance, are just one member of a family of particles called leptons, while protons and neutrons are part of a family called hadrons, of which hundreds are known. This is beyond the scope of this book but any interested layman might look for *The Discovery of Subatomic Particles* by Nobel laureate Steven Weinberg. It is an easily read exploration of a complex subject.

The special property of electrons, protons, and neutrons that makes them the building blocks of ordinary matter is their relative stability. Electrons are considered truly elemental and are believed to be absolutely stable, unlike other particles with very short lifetimes. Protons and neutrons, when bound within an atom's nucleus, live for ten to thirty years.

Energy and the Electromagnetic Spectrum

Energy is the dynamic of one elemental state's being changed into another. (And each state also has its own energy level, too.) Energy takes many forms, such as light, heat, sound, electricity, and nuclear activity, to name but a few. Radiation is the term used to describe energy in motion radiating away from its source. It is expressed in wavelengths calculated along what is called the electromagnetic spectrum. There are many wavelength magnitudes along the spectrum, which includes everything from the earth's own natural pulsations, to electric power, to visible light, to cosmic events.

The electromagnetic spectrum is divided into ionizing and nonionizing radiation. Ionizing radiation consists of very short wavelengths (like X-rays), which have enough power to knock electrons off their nuclear orbits and therefore can cause perma-

nent damage on the cellular level, such as cancer or genetic mutations. Nonionizing radiation consists of longer wavelengths with
generally less power and is considered incapable of knocking
electrons off their orbit around the nucleus. Energy is directly
related to the inverse of wavelength. The longer the wave, the less
energy it creates; the shorter the wave, the more energy it can
create.

Although less powerful, nonionizing radiation is capable of
causing a host of biological effects, so it is far from harmless. It is
the nonionizing bands that are the main focus of this book.

The dividing line between ionizing and nonionizing radiation
is around visible light. The problem is that visible light covers a
wide portion of the spectrum. (Each color, for instance, is a different wavelength.) So no one can say precisely where one form of
radiation leaves off and the other begins, but it is thought to be in
the lower ultraviolet ranges, or possibly below.

Some common consumer products have complex electromagnetic fields. A television screen or the cathode-ray tube on a
computer monitor, for instance, plugs into a wall outlet, thereby
using electricity in the 50–60 Hz area, then utilizes a broad band
of wavelengths that extend all the way up through the visible-light
spectrum. At the top ends of the screen's field, the ionizing band
may well be crossed into.

How to Read the Electromagnetic Spectrum

A common way of depicting the spectrum is shown in this
illustration entitled "Electromagnetic Spectrum." The numbers
on the sides are just units of measurement, and readers who are
comfortable with that can skip this brief explanation of what they
mean.

Scientists have figured out a handy, economical way to deal
with very large or very small amounts of things. Atoms, electrons,
and wavelengths certainly fit the bill. This system makes saying a

Frequency (in hertz)	Type of Radiation	Examples of Sources	Wavelength (in meters)
	Cosmic rays		— 10¹⁵
10²³—		Sun	— 10¹⁴
10²²—			
		Nuclear material,	— 10¹³
10²¹—		diagnostic and	
		therapeutic X-ray	— 10¹²
10²⁰—	Gamma rays	equipment	
			— 10⁻¹¹
10¹⁹—			
			— 10⁻¹⁰
10¹⁸—	X-Rays		
		Welding equipment	— 10⁻⁹
10¹⁷—		mercury-vapor lamps,	
		black light (UV) devices,	— 10⁻⁸
10¹⁶—	Ultraviolet radiation	fluorescent and incandescent lights	
10¹⁵—	Visible light	White light devices	— 10⁻⁷
10¹⁴—		Arc processes,	— 10⁻⁶
		lasers, hot furnaces,	— 10⁻⁵
10¹³—	Infrared radiation	molten metals/glass,	
		alarm systems,	— 10⁻⁴
10¹²—		motion detectors	
			— 10⁻³
10¹¹—	EHF		
		Cellular phones,	— 10⁻²
10¹⁰—	SHF Microwaves	microwave ovens,	
		radar, medical	— 10⁻¹
10⁹— (GHz)	UHF	diathermy equipment,	
		ultrasound, smoke	— 1
10⁸—	VHF	detectors, MRI	
10⁷—	HF Short wave		— 10
		Communications	
10⁶— (MHz)	MF Radio Medium	equipment, CB	— 10²
	waves wave	radios, AM/FM	
10⁵—	LF Long	transmitters	— 10³
	wave		
10⁴—			— 10⁴
10³— (KHz)			— 10⁵
			— 10⁶
10²—	Extremely low	Power-generating	
	frequency (ELF)	equipment, 60 Hz	— 10⁷
10¹— (Hz)		appliances like stoves, hair dryers, etc	

Ionizing (10¹⁵ and above) / *Nonionizing* (below 10¹⁵)

Electromagnetic Spectrum

Power of 10	American Term	Prefix
10^1	ten	deka
10^2	hundred	hecto
10^3	thousand	kilo
10^6	million	mega
10^9	billion	giga
10^{12}	trillion	tera
10^{-1}	tenth	deci
10^{-2}	hundredth	centi
10^{-3}	thousandth	milli
10^{-6}	millionth	micro
10^{-9}	billionth	nano
10^{-12}	trillionth	pico
10^{-15}	quadrillionth	femto

jawbreaker like quadrillion unnecessary, and simplifies mathematics, too.

The system commonly used employs powers of 10. The notation 10^1 is just 10; 10^2 is the notation for two tens multiplied (10×10, or 100); 10^3 is the notation for three tens multiplied ($10 \times 10 \times 10$, or 1,000); and so forth. And the system works in the opposite direction for negative numbers.

The general mathematical rule is this: in multiplying powers of 10, add the powers; in dividing, subtract them. For example, $10^{22} \times 10^4$ equals 10^{26}; 10^{25} divided by 10^{18} equals 10^7. To handle numbers that are not in units of ten, they can be combined like this: 3×10^2 would equal 300.

The numbers on the electromagnetic spectrum that look like this represent the number and size of wavelengths per second. Wavelengths are measured in hertz (Hz), named after the German physics professor Heinrich Hertz, who built on the forgotten work of the Italian physician Galvani by demonstrating in the mid-1800s that "action at a distance" was possible without a connecting wire. Hertz also showed that energy moved in wavelengths that oscillated along the electromagnetic spectrum. Frequency is the term used to describe the fluctuation (or oscillation) per second of a wavelength. For instance, one fluctuation per second would be a 1-hertz frequency field. Hertz's discoveries led directly to Guglielmo Marconi's invention of the wireless telegraph in 1896, which would forever change the world.

As the illustration shows, the wavelength of electricity is long in comparison to those of cosmic rays and X-rays, which are very short. A frequency of 1 hertz has a wavelength measuring in the millions of miles; a 1-million-hertz frequency (1 Megahertz, or 1 MHz) has a wavelength several hundred feet long; and a 100-million-hertz frequency (100 MHz) is about six feet long.

All wavelengths travel at the speed of light, although they differ in size and strength. Light is the only visible form of electromagnetic radiation. It was Albert Einstein who theorized that nothing travels faster than the speed of light—a theory still accepted by physicists, though some say Einstein himself had doubts about it toward the end of his life.

Electromagnetic Fields: What They Are

Electromagnetic field is the term used to describe electric and magnetic fields. But each has different properties that need to be understood separately before they can be understood together, even though they are considered inseparable, as one will create the other.

In general, a field is defined as a space in which energy exists. Actually a field is a concept used to describe the influence of something on its surrounding area. A heat field, for example, is what exists around a crackling fire or a burning light bulb.

Electric Fields

Our most common experience with electric fields is with static electricity, like that produced in a dry indoor atmosphere when we walk in stocking feet across a carpet, touch a metal object, and get a shock.

Static electricity is called that because it comes from electrons that are moved from one place to another rather than flowing in a line of current. In a static field, the positive and negative charges have become separated over some distance. The large separation of the charges can produce a pretty good static electric field as you move across the carpet; you feel it when the hair on your hands or arms stands up as you near a grounded conductive object (of metal, say, or some forms of plastic). If you touch the conductive object, a shock will occur as the electric current flows through, returning the separated electric charges back to a neutral state. If you immediately touch the object again, you will not get another shock because the charges have become neutral. Positive and negative charges attract each other; like charges repel each other.

Electric fields exist whenever electric charges are present, that is, whenever electricity or electrical equipment is in use. Electric fields are produced when the positive and negative charges are separated; the potential difference in this separation is called the voltage. Voltage is a potential—like the water in a garden hose with the nozzle turned off. The strength of electric fields increases or decreases according to the voltage—the potential ability within the hose.

At any given time, the voltage increases as more of the charges are separated. The longer you drag your feet across the carpet, the greater the separation between the charges, the more the potential difference between them, the greater the possible release of energy when they come together again. An electric field can be strengthened by increasing the potential difference or by moving the opposing charges closer together.

When a conducting path, such as a wire, is provided between the area of the separated charges, the charges will flow between the two regions in the form of electrons or ions (which are atoms with an unequal number of positive or negative electrical charges).

Electric Current

The flow of electrons is called electric current, and it is measured in amperes (A), after the French mathematics professor André Marie Ampère who, in 1820, discovered that parallel wires attracted or repelled each other if they carried electric current flowing in the same or the opposite direction, respectively. Ampère also discovered that all magnetism is electrically based, that it is the minute electrical properties circulating in the particles of a lodestone, for example, that impart its magnetic abilities.

The electric charge itself is measured in coulombs (C.), named after Charles Coulomb. Current is simply the number of coulombs of charge that flow through a given region per second. Electric fields are usually known as vector fields, because they specify both a magnitude and a direction for each point in space that a positively charged ion would travel if placed there.

Some things, like air, act as insulators (or dielectrics), across which current will not usually flow. However, given enough voltage, there is no effective insulation—against lightning, for example. And high humidity inhibits air's insulating properties. Also, frequencies above electricity, like radio signals, will easily flow through space.

The illustration called "AC and DC Current" shows the two kinds of electric current. Direct current (DC) is the steady flow of electrons in one direction. Alternating current (AC) is an electron flow that changes strength and alters direction within a certain cycle; the AC field collapses and reappears with its poles reversed every time the current changes direction.

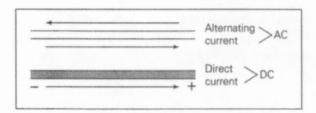

AC and DC Current

Batteries are a form of DC power. AC power can be understood with the standard American 60-hertz cycle, in which electric current changes direction and strength 120 times per second, making 60 complete cycles per second. In Europe, the cycle is based on 50 hertz, meaning that the current changes direction and strength 100 times per second, or 50 complete cycles per second.

Direct current creates a steady magnetic field. But with alternating current, each time the direction of the electrons is reversed, or flipped, a powerful magnetic field is created that fluctuates at the same frequency. Some of the earliest electrical-power grids, created by Thomas Edison around 1882, were DC based. A few antiquated DC systems still exist in New York City, including the entire subway system. In 1893, an extraordinary genius named Nikola Tesla created the first AC power system when he lit the Chicago World's Fair. Within two years, he had ushered in modern electrical engineering when he harnessed the power of Niagara Falls and created a system to transport that power.

Tesla's discovery of alternating current allowed electricity to travel over long distances, the way we see today on high-tension wires. Edison's DC system could only travel relatively short distances because it was difficult to convert between the voltages needed for high-tension transmission and low-distribution end use.

It would be possible to eliminate many of the low-current magnetic fields in our midst today by going back to a DC transmission system. Some utility companies are returning to high-voltage DC transmission lines, but only for systems spanning great distances whose power does not need to be tapped off along the way. One such system was activated in 1970 from the Pacific Northwest to the Pacific Southwest. The advantages and disadvantages of each will be discussed in more detail later on.

Magnetic Fields

The most common experience we have with magnets is associated with those horseshoe-shaped iron magnets that many of us played with as children. But whether it was a flat bar magnet or one bent into a U shape, the physical dynamics are the same.

A fixed magnet's field is created by the spinning of electrons around the nuclei in the iron itself. For this to occur, the atoms must be aligned in the same direction so that the individual fields within each atom combine into one big magnetic field. It's easy to undo the alignment—just drop a magnet on a very hard surface and see how it loses its magnetic potential when the atoms become scrambled by the force of the fall.

All magnets have two poles, north and south. Opposite poles attract; like poles repel, as the illustration called "Magnetic Fields around a Bar Magnet" shows. This means that magnetic fields have a direction (vector), as well as strength, like electric fields. The strength of the magnetic field is determined by the amount of current that is flowing. To use the hose analogy again, is the nozzle fully turned on or only partially? Is it a small garden hose or a powerful fire-fighting hose that can reach hundreds of feet in the air?

A magnetic field is produced whenever there is electrical current flowing in a conductor or wire. All magnetic fields run

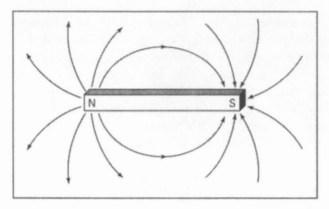

Magnetic Fields around a Bar Magnet

perpendicular to electrical current. In other words, while electric current runs in straight lines, magnetic fields surround the line in circular fashion. (See illustration called "Electric/Magnetic Fields.") When the electricity is turned off, there is no magnetic field, although an electric field will still exist to some extent in any wire or conductor.

If the electric current is DC, the magnetic field will be steady,

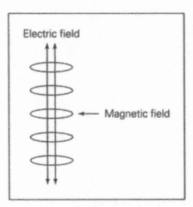

Electric/Magnetic Fields

like that of a permanent magnet. If the electric current is AC, the magnetic field will fluctuate at the same frequency as the electric field—that is, it becomes an electromagnetic field, because it contains both fields. (In most cases, the two fields are inseparable.) The "Lamp Off/Lamp On" illustration shows how this happens.

The term "electromagnetic field" is really a combination that takes off into space at radio frequencies (and higher), although we commonly use it today to include power-line frequencies. Normally the electric and magnetic fields can be calculated from each other, but at power-line frequencies an electric field can exist without a magnetic field. These are called near fields, and they do not radiate away, as the higher frequencies do.

All electromagnetic fields are expressed in wavelengths and move outward at the speed of light—approximately 186,000 miles per second.

Both electric and magnetic fields decrease rapidly with distance from the source. This is a point that will be emphasized over and over throughout this book.

Electric power, which is the product of voltage and current, is measured in watts (after James Watt). Electric fields are measured in volts per meter (V/m) or, when large amounts are described such as those used in high-tension transmission, kilovolts (1,000 volts) per meter (kV/m) is used. The higher the voltage, the lower the losses for equal amounts of energy being transmitted.

Magnetic fields are measured in teslas (T), named after Nikola Tesla, and in gauss (G), named after Karl Friedrich Gauss, the nineteenth-century German pioneer in magnetism. Because the units are large and interchangeable, measurements are often expressed in microtesla (μT) and milligauss (mG). One tesla (1T) equals 10,000 gauss; one gauss (1 G) equals 1,000 milligauss; one microtesla (1 μT) equals 10 milligauss (10 mG).

Some electric and magnetic fields are easy to measure with inexpensive meters, which are discussed in a later chapter.

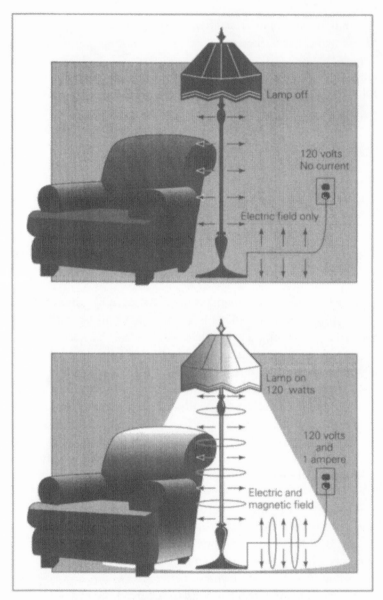

Lamp Off/Lamp On

Not All Fields Are Alike

Not all frequencies, waves, or fields are alike, nor is their impact on living organisms the same. As we have said, the higher the frequency on the electromagnetic spectrum, the shorter the wavelength. The shorter the wavelength, the more readily it is absorbed by living things. And the more readily absorbed, the more likely are biological interactions, although absorption is not the only factor that can create biological effects.

For instance, an object the size of an average human being absorbs about one trillion times less energy from the long waves of a 60-hertz power-line field than from a 60-megahertz television frequency field of equivalent power intensity. Also, the lower the power and the longer the wavelength, the less that power is radiated away. High-frequency radio and TV transmission towers radiate significant amounts of energy, whereas power lines radiate far less. (High-voltage DC transmission lines radiate less than AC lines.) This, however, does not mean that significant biological alterations do not occur at lower frequencies. In fact, our bodies may be more attuned to them in some ways than to the higher frequencies.

There are a variety of wave forms as well. Curved sine waves, for example, occur around AC power frequencies. Waves generated by electronic equipment often have several forms, including rounded sine waves and pulsed waves that can have a sharp saw-tooth or square shape. There is some evidence that pulsed saw-tooth forms are more harmful.

How Electromagnetic Fields Affect the Body

Although electric and magnetic fields are often dependent on each other, they act quite differently in some critical ways, especially in their effect on living organisms. In general, when electric and magnetic fields interact with living things, they

separate, or uncouple from each other, then recouple and interact as different forces with the organism's inate fields.

Most low-level electric fields are considered superficial in
their ability to penetrate matter, including the human anatomy,
because the human cell's membrane is a relatively good dielectric
(insulator), up to a point. Also, many objects act as shields against
electric fields, including trees, houses, rocks, and so on. But magnetic fields can penetrate just about anything that does not have
a high iron content, including buildings and the human anatomy.
Most of the concern about EMF bioeffects centers around the
magnetic component, not the electric one.

Matter falls into one of two categories in relation to electromagnetic fields. Everything is categorized as either a conductor —
it will allow electricity to flow through it — or as an insulator — it
will not transmit electricity. The human anatomy is a pretty efficient conductor, which means that the body can serve as an antenna in the presence of a strong electromagnetic field.

In fact, the human body has a higher conductivity than air,
allowing electric current to flow efficiently, though superficially,
through the skin. Unless the body forms a ground with a strong
external current, in which case shock or even electrocution will
occur, people usually do not even feel most contact currents.

But the human body will take on whatever field it is exposed
to. Each time you touch a small electric appliance operating at 60
hertz, those same 60-hertz fields will be set up in your body as
well. (Such small contact currents are usually not of much concern, but a process called resonance is. This will be discussed
further in Chapter 8.)

Magnetic fields are another matter. The human body has a
magnetic permeability almost identical to that of air, and therefore magnetic fields are deeply absorbed in their entirety. Alternating magnetic fields induce electric fields and currents inside
living things, and a range of biological effects is possible. This will
be discussed in more detail later, but it is important to note that
current pathways caused by magnetic fields (called eddy cur-

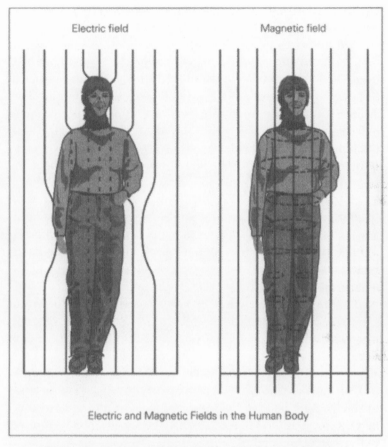

Electric field

Magnetic field

Electric and Magnetic Fields in the Human Body

*Electric and magnetic fields induce weak electrical currents
in humans and animals. This is what would occur when standing
under a power line. The dotted lines show the direction of induced
current flow. The solid lines represent invisible field lines.*

rents) are different than those caused by electric fields. The illustration "Electric and Magnetic Fields in the Human Body" shows
what occurs in someone standing under a power line. The mag-
netically induced body current flows primarily in peripheral loops

at right angles to the field, with the current at the center of the body near zero, whereas electrically induced current flows in straight lines.

This is important to remember because an externally created magnetic field has the ability to affect the body's natural electrical balance. In other words, external fields can induce internal fields. Magnetic fields can also magnetize certain things with magnetic potential, such as iron, which is stored almost exclusively within the blood. Magnetic fields can also exert a force on the blood's ions or electrolytes. And who knows what occurs with the many trace minerals like copper in the body?

Although we don't normally think of ourselves this way, on the most elemental level we are a huge swirl of positive and negative electrical particles in dynamic equilibrium with each other. These form the many molecules, tissues, and mineral compounds in the body. In a way, we are one big moment-to-moment exchange of interacting electrons, with our natural electric and magnetic properties fueling life's dynamo. The introduction of external electromagnetic fields into this already complex anatomical structure warrants far more careful study than it has had to date.

Although the focus is largely on genetics today, molecular medicine and molecular biology are two promising areas for study of the health effects of electromagnetic fields on the human organism. A newly formed scientific specialty called bioelectromagnetics is another promising area.

Complexities of the Subject

Bioeffects from externally generated electromagnetic fields encompass many variables. These include the body's size, shape, tissue type, orientation toward and distance from the generating source, the duration of exposure, the size of the wavelength, and the power with which it is being generated. These variables make a detailed understanding of cause-effect ratios extremely complex, if not outright impossible — and this is on top of the inherent complexities of the properties of electromagnetism itself.

Different tissue types absorb wavelengths and frequencies very differently. This is true of different human tissue structures (skin is different from the eye or the liver, for instance), and it is also true of other species. The human anatomy most efficiently absorbs radiation of around 87 megahertz in the FM radio band and resonates (reaches maximum absorption) at between 30 and 300 megahertz, which encompasses frequencies spanning FM radio and very high frequency (VHF) television broadcasting. A fruit fly, on the other hand, absorbs the microwave band most efficiently; a very small dose of microwaves is quickly fatal.

Those Confusing Measurements and Terms

The measurements for radiation can be extremely confusing to the layperson. A commonly used term is power density, which is measured in milliwatts per square meter (mW/m^2). Power density, a vector quantity that describes the rate at which energy is transmitted through a given area, can be a useful tool in bioeffects studies concerned with the amount of energy input in living organisms.

But to use power density alone is misleading. For one thing, the actual field strength may be different than the power-density measurement. In the real world, fields react with each other in complicated ways. For instance, near radio antennas the fields closest to the transmission towers often collapse in on themselves, so that a power-density prediction at various distances would be inaccurate according to the usual mathematical formula.

Also, in real-world reactive fields, no fixed relationship exists in all cases for electric and magnetic field components. What this means is that a person or animal acting as a conductor within a given power-density calculation might absorb and store energy from both electric and magnetic properties differently than the power-density measurement could predict. The effects from a low field could be absorbed and stored in a much higher capacity.

For this and other reasons, a more meaningful measurement has been devised called the specific absorption rate (SAR). This measurement presumes the direct effect on an organism rather than the application of a spatial mathematical theory. SAR is the measurement of the power absorbed by whomever or whatever is being studied. But again, because living organisms are complex, it is difficult to accurately predict SARs. A whole-body SAR would be different than the SAR of an individual organ, like the brain. The measurement is especially useful in studies concerning thermal effects (the heating of tissue), because the absorption of electromagnetic energy can be compared to the rate at which the body produces energy through normal metabolism. Excess absorption will produce heat, and exposure limits can be set below this level, using an "average" human model. This works in theory, but reality is another matter. Women and children are not factored in, and nonthermal effects are left out altogether. But the SAR concept is a useful one, nevertheless.

The bottom line, however, is that as long as the basic mechanism for adverse effects from electromagnetic fields remains largely unknown (although various possibilities have been hypothesized), we cannot say what measurement of internal exposure is most relevant. In the face of this, government agencies and manufacturers might well adhere to a principle that has generally been adopted for exposure to ionizing radiation and accept it for the nonionizing bands as well. The principle is called ALARA, which stands for "as low as reasonably achievable." The mitigation of exposures across the board should be factored into every product and every decision.

Comprehending the Invisible: How Things Work

Many people think that physics and electromagnetism are theoretical constructs with little meaning for real life, probably because the scientific concepts are so abstract that it is hard for us to relate to them. The principles may be abstract, but they function all around us in a quite tangible fashion. Any ball game, for instance, is physics in motion, and many aspects of electromagnetism can be understood through such a model.

Imagine that power lines are basketballs; radio waves are softballs; microwaves are hardballs; and infrared radiation is golf balls. If you are standing in the path of any one of these projectiles, you will experience a different response to each, depending on which ball it is (the wave), how forcefully it is coming at you (voltage and current), and where it actually contacts your body. A basketball will give you a good thump; a softball will leave a black-and-blue mark; a hardball might cause serious damage; and a golf ball could penetrate body tissue. (If we continue the analogy into the ionizing bands, X-rays might be comparable to a game of darts.) Also, a ball that hit you in the padded area of the thigh would have a quite different effect than one hitting you in the temple.

A baseball field is the vector of the game: how hard a ball is hit determines the distance and strength of the play. The force of the pitcher's arm is like voltage, and the number of balls thrown and hits made are like the coulombs of current.

Unfortunately, the ball-game analogy is an imperfect one, because wavelengths are not round and they often move through objects. Sometimes the analogy of different-sized ships on the ocean all moving at the same speed is useful, but even that has its limitations. Nothing is quite like wavelengths, after all.

The picture becomes still more complex with such technologies as radio, TV, sensing devices, and computers, because other factors come into play. Radio transmission involves an electrical power source which is sent to a metal transmitting mast, where the atom's electrons are excited into the higher radio frequencies. This employs an oscillator, a device that alternates the electrons' polarity within a given time frame, as well as the process called modulation.

Modulation is the key to transmitting information over a distance. It adds another signal, which rises and falls at a set interval and travels on what is called a carrier signal. The carrier signal is said to be modulated when it is combined with another signal, such as a sound or a visual frequency. The modulated frequency is projected toward a receiver, which detects the sound or visual signal from the carrier wave, extracts it, and is then able to reproduce that sound or image alone. See the illustration, "How Modulation Works." Fiber optics is an example of modulation; it combines light and sound frequencies together and fires them along a glass cable. Radio and TV transmission is another example of modulation.

To return to the baseball analogy, imagine that the pitcher wants to send the catcher a piece of chewing gum during a game. The gum by itself is too lightweight to be thrown that far. So the pitcher sticks it onto a ball and throws them together. But first he swings his arm around again and again to oscillate, or rev up, the throw. The catcher (receiver) can then remove the gum from the ball.

Modulation is an important concept for consumers to understand. Often when industry takes measurements of various output sources, they measure unmodulated signals. Critics say this is an effort to minimize readings in safety tests. It's one area where consumers should insist on tightening up the measurements, especially around radio, TV, and cellular-phone towers.

There are several different kinds of radio waves, transmitters, antennas, and receivers, all of which will be discussed in a subsequent chapter. For now, what is important to understand is that the dynamics of what is needed to transmit energy and what is

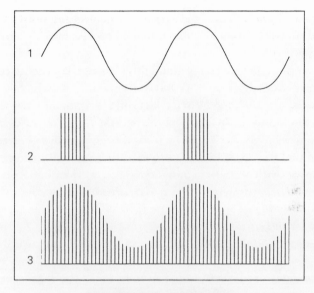

How Modulation Works

Line 1. Primary signal.
Line 2. Pulsed microwave or radio-frequency signal
that is turned on at specific fractions of a second.
Line 3. Amplitude modulation as the signals
combine and smoothly oscillate together.

required to receive it are not the same. Transmission always requires more energy, and therefore exposures to electromagnetic fields will always be higher near transmission devices than near receiving ones.

Satellite up-links (which send information to satellites in orbit around the earth) have much stronger electromagnetic fields than do the satellite-receiver dishes that people put in their backyards, which do not radiate but, rather, collect the signals. Radar installations send out a strong, constant scanning signal. Radio towers generate much stronger fields than a radio itself, which is just a receiver. Generated signals are present all the time, ready for antennas to pick them up; radio waves don't cease to exist just because we turn off the radio. And some devices are both receivers and transmitters, like walkie-talkies, cordless phones, and car phones.

Television broadcasting is slightly different. Although TV broadcast towers can emit 50,000 watts of energy—which is a lot of power—the technology of a television set, while nowhere near that power output, can also generate strong multifrequency electromagnetic fields. A TV set functions through the use of electron guns aimed from inside the set toward the screen, or directly at the viewer, with the phosphorus coating and the glass over the screen the only things in the way. The area closest to the set can emit a fairly high electromagnetic field. A viewer should always sit at least six feet away from the screen.

Sensors and detectors work in various fashions and utilize a range of electromagnetic frequencies. The scanning devices at supermarkets are lasers in the infrared range; the Breathalyzers that measure alcohol levels also use the infrared range, as do motion detectors like automatic door openers. Smoke detectors come in different varieties that use electrical frequencies, light frequencies, and some ionizing bands as well. Computers are multifrequency devices, depending on the type of screen (cathode-ray tube, radioactive gas plasma, or liquid-crystal display).

All of these are artificially created frequencies. None of them existed in the natural world before the twentieth century.

Chapter 4

NATURE'S
ELECTROMAGNETIC FIELDS

T HE EARTH ITSELF is a gigantic complex dipole magnet—not unlike the common bar magnet, with its north and south poles—with a spinning core of molten iron and nickel thousands of miles beneath its surface. Micropulsations in the 10-hertz range emanate constantly from the planet's core.

Scientists once thought of this as an interesting but innocuous phenomenon. But science is increasingly coming to understand that all of life, including those most basic daily rhythms of waking and sleeping, are affected by our interactions with this intricate natural magnetic environment. Many now think that all living things are in a subtle—and sometimes not so subtle—harmony with it, that it influences everything from the migration patterns of birds down to the very cell division (mitosis) that takes place in the human body.

The Geomagnetic Envelope
and the Magnetosphere

The earth's natural electromagnetic environment is created by an elaborate interrelationship between the magnetic core, the charged gases of the ionosphere, and the enormous power of the solar winds constantly being given off by the sun.

Solar winds consist of high-energy ionizing atomic particles that travel at a tremendous speed millions of miles through space and bombard the outer layers of the earth's magnetic field. The sun also gives off vast amounts of ionizing radiation in many frequencies, including X-rays, infrared, ultraviolet, gamma, and cosmic rays. A wondrous and complex protective layer called the magnetosphere protects us from this otherwise deadly barrage, which on the one hand makes life possible but on the other hand would kill all life instantly, if given unlimited access.

The magnetosphere itself has several components, sometimes diverting the sun's rays away from the earth and sometimes absorbing them. The bow-shock region is the name given to the area where the solar winds impact the outer magnetic fields of the earth on the side facing the sun. The opposite side of the magnetosphere extends into a long magnetotail reaching away from the earth and far into space.

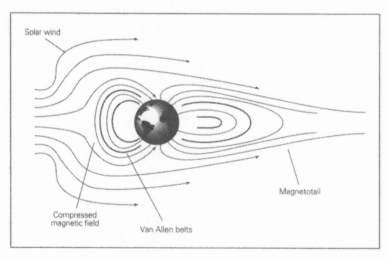

Our complex magnetosphere formed by the interaction between the solar winds and the Earth's natural magnetic fields.

As these two forces collide, the earth's magnetic field is compressed to match that of the solar wind. The area of the bowshock region that absorbs much of the solar impact is called the Van Allen Belts (named after the American physicist James Van Allen), which act as ducts to trap the high-energy particles and spiral them back and forth between the north and south poles.

The magnetosphere remains stationary in space, with one side always facing the sun as the earth rotates on its twenty-four-hour day-night cycle. This means that any given place on earth has a constantly changing magnetic field. It is the rise and fall in the strength of this natural magnetic field that creates all biological rhythms—the circadian rhythms, in humans.

The interaction of energy from the solar winds and the earth's magnetic field also creates large natural electrical currents that can reach billions of watts. In addition, it creates ionizing radiation as well as nonionizing radiation in various frequencies, including extremely low frequency (ELF) waves between 0 and 100 cycles per second and waves in the very low frequency (VLF) range of 100 to 1,000 cycles per second.

The sun, too, is a varying heavenly body. The power of its solar winds and storms changes tremendously, and it displays an eleven-year cycle marked by huge eruptions of energy on the sun's surface, called solar flares. Solar flares, which are marked by a steep increase in energetic particles within the solar winds, cause solar storms. They in turn flood the magnetosphere with X-rays and streams of protons, gamma rays, and radio waves that cause magnetic storms on earth.

These magnetic storms occur in certain layers of the earth's atmosphere, like the ionosphere. They can cause major disruptions in radio and TV signals, as well as increases in the voltage in telephone and electrical transmission lines that are sometimes powerful enough to cause equipment damage. Some studies have also correlated magnetic storms with an increase in admissions to psychiatric hospitals.

The beautiful nocturnal lights of the aurora borealis are

another effect of solar storms; particles from the solar winds enter the magnetosphere envelope at either of the polar regions, where the Van Allen Belts become narrow, penetrate into the earth's upper atmosphere, and interact with gas molecules, which then produce those soft celestial lights. Unfortunately, now that there is so much earth-generated light pollution in the temperate zones, especially from orange sodium-vapor lights, fewer and fewer of us will be able to experience the aurora borealis without traveling to one of the polar regions. Whole generations of inner-city dwellers and suburbanites will never see this wonder, except perhaps as a fortuitous by-product of an electrical blackout.

More ominous, however, is the fact that the artificial radiation in the ELF and VLF ranges (from power lines and some radio signals) may be capable of entering the magnetosphere and changing its structure. As energy bounces back and forth from pole to pole along the magnetic ducts, a resonance factor may be amplifying some of these low-level energy sources.

This has profound implications for global weather patterns. There has been a 25 percent increase in thunderstorms in North America alone between 1930 and 1975, as compared to the period between 1900 and 1930, as well as drastic increases in flooding in some regions and drought in others within the last few decades. At a conference on electromagnetic compatibility, held in Zurich, Switzerland, in 1983, one researcher, K. Bullough, suggested that this was due to increased energy levels in the upper atmosphere. A phenomenon called power-line harmonic resonance may have created a permanent duct from the magnetosphere into the higher altitudes capable of releasing a continuous flow of ions and energy over the North American continent, according to measurements taken from weather satellites.

In addition, Dr. Robert Becker notes, in *The Body Electric*, that radio and microwave energy is also known to resonate in the magnetosphere and to interact with particles in the Van Allen Belts, which can create a fallout of charged particles that serve as nuclei for raindrops. Specific areas of ion precipitation have been

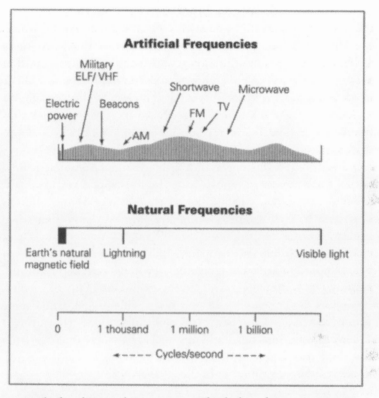

*Artifical and Natural Frequencies: The darkened areas represent
some of the earth's natural electromagnetic environment. Shaded areas
represent the artificially created frequencies of man in the nonionizing
band. The naturally occurring empty spaces of the electromagnetic
spectrum have all been filled in with our technologies.*

matched to certain radio-station transmitters through patterns
of charged particles that follow weather movement in a general
easterly direction.

What all this means is uncertain, but the implications are
serious and deserving of immediate attention, as these artificial
energy sources fill in more and more of the electromagnetic spectrum. (See illustration of "Artificial and Natural Frequencies".)

Our Complex Electromagnetic Orb

Unlike the simple bar magnet, the earth's electromagnetic fields have incredibly subtle, complex interrelating aspects that create a wondrous orb to cradle, sustain, and influence all of the earth's biological systems. A great deal about the subject remains a mystery, and much of what geophysicists have postulated remains theoretical, as in fact do electromagnetic waves themselves. We are still in our infancy regarding this subject.

One thing we do know is that the earth's electromagnetic field varies from month to month with the cycles of the moon as it revolves around the earth, and that there is a yearly variation as the earth in turn rotates around the sun. We know that some cycles occur for centuries and then change—no one knows why. Also, the earth has the disturbing habit of reversing its polarity every few million years. Each reversal, which can take a thousand years or so to complete, results in the extinction of major species, as well as the appearance of new ones. This probably had some role in the disappearance of the dinosaurs, and perhaps thrust Homo sapiens into becoming an evolved species separate from the great apes. There are some indications that we are in the process of undergoing another of these reversals.

The interplay between the earth's surface and the ionosphere forms an electrodynamic resonating cavity from which come the 10-hertz micropulsations in the extremely low frequency ranges (numbering from approximately 25 pulsations per second to perhaps 1 every ten seconds). Large direct currents move continuously within the ionosphere, as well as enormous direct currents within the earth itself, all of which generate their own electromagnetic fields. Also, the earth's surface and the ionosphere act in tandem, like a big condenser with two charged plates, producing electrostatic fields in the hundreds of thousands of volts per foot.

The manner in which this affects the surface of the earth,

and therefore our relationship to it, is curious, to say the least. For instance, we tend to think of a thunderstorm as a localized event that rattles the windows and sends the cat hunkering toward the darkest corner of the closet. But each thunderstorm has global effects. Every flash of lightning releases a huge burst of radio-frequency energy measuring in the kilohertz ranges. This energy travels back and forth parallel to the earth's magnetic-field lines between the north and south poles, over and over, before gradually fading away.

The soil continually emits into the air electrically charged particles called ions, which are either positively or negatively charged. On a clear sunny day, the earth has a negative charge and the atmosphere has a positive charge, so that electrons stream skyward from the soil and from plant life. But during stormy weather, the polarities are reversed, with the earth becoming positive and the bottom of the cloud base negative.

At any given time, perhaps three or four thousand electrical storms are raging all over the globe. A dynamic equilibrium occurs between the ions being released in clear-weather areas and those in stormy regions so that the earth's electrical gradients are maintained.

Along with this constant flow of electricity in many directions, the natural background voltage increases with altitude. Between the lowest layers of the ionosphere and the earth's surface, there are about 360,000 volts; from the top of the Empire State Building to the sidewalks at its base, there are approximately 40,000 volts, and between the head of a six-foot man and the sidewalk, approximately 200 volts. At the earth's surface, the voltage is approximately 100. This sounds truly frightening, but in fact hardly any shock potential exists because there is no current; it is just background voltage and static DC fields. We know very little about this huge reservoir of natural power—how it functions or what governs it—but it gives rise to some interesting speculation about the implications for those who live or work in

high-rise structures or who spend time at high altitudes, like airline personnel—all of whom experience increased exposures
from the artificial EMFs generated from Earth as well.

The Earth's Magnetic Grid, Geopathic Stress, and the Dowsing Phenomenon

The earth is composed of a magnetic-grid-like system somewhat like the segments of an orange with intersecting lines. These
grid lines are called ley lines, and they can vary in intensity from
place to place, based on variations in gravitational force, the presence or absence of large mineral deposits like quartz, and the
presence of underground streams or large aquifers, all of which
can alter the electromagnetic background on the earth's surface.
Some European architects take ley-line fields into account in the
design and placement of buildings, as well as in what construction materials they choose. Some preliminary interest in this is
being shown in the United States, too.

All living organisms are "entrained" by the earth's planetary
fields, which vary in strength and which can have both beneficial
and detrimental effects. European research is far ahead of that
in the United States regarding the impact on human health of
abnormal fields associated with particular geographical hot spots.
Some studies have found a statistically significant increase in
cancer in people who live above large active underground aquifers, for instance. English and German studies have observed that
living in regions of geopathic stress not only contributes to the
onset of illness but may hamper effective treatment, too.

Although it is unlikely that abnormal geological fields alone
can cause cancer, these studies do add an interesting dimension
to various observations of a higher cancer incidence in certain
regions, cancer clusters in some work environments, and tendencies for particular cancers to run in families. Dr. Richard Gerber,
in his book *Vibrational Medicine,* postulates that the geopathic

field probably acts in concert with a variety of other predisposing factors, including diet, genetics, environmental carcinogens, viruses, abnormal electromagnetic radiation exposure, and other factors that can affect general vitality and immune competence.

So how do you find out if you live in such a stress area? One way is to hire an old-fashioned dowser. Many rural communities still have a dowser or two around. A good place to start is with the older well-drilling companies which at one time relied almost exclusively on dowsers before attempting any well installation. Or contact the American Society of Dowsers, Inc., at P.O. Box 24, Danville, VT 05828-0024; phone: 802-684-3417 for information about dowsers in your area. (Be forewarned, however, that different dowsers use their own individual "maps" and they can sometimes contradict each other.)

Dowsing is a millennia-old tradition worldwide. References to it are found in the history of China, India, Egypt, Persia, Media, Etrusca, Greece, and Rome. Although often considered witchcraft in former times and quackery in our own time, dowsing is not without its scientific proponents, including many geologists and some in the U.S. military.

One of the tools dowsers use is a two-pronged wood or metal fork, with one prong held in each hand and a single extension facing away from the dowser, upon which enough force is exerted so that it is pulled downward when it advances over water, metals, or a mineral deposit. Another method uses an L-shaped tool, and another employs a pendulum suspended on a string between two poles that the dowser holds between both hands while walking along.

Far from being witchcraft or quackery, dowsing appears to have the ability to tap into the earth's DC magnetic fields and locate variations in them, through what might be considered a primitive antenna—the dowsing rod. It is thought that the dowser's body serves as a conduit between the earth's DC magnetic fields and the "divining" tools, as they used to be called.

In the United States, Dr. Zaboj V. Harvalik, a retired physicist and former scientific adviser to the U.S. Army's Advanced Material Concepts Agency, has investigated the dowsing phenomenon and helped establish a physical theory to explain it. Dr. Harvalik's research found that dowsers react in varying degrees to polarized electromagnetic radiation, artificial alternating magnetic fields ranging in frequency between one and a million cycles per second, and to DC magnetic fields. Dr. Harvalik found evidence that dowsers' bodies sense magnetic-field gradients from many things such as underground pipes, tunnels, wires, and geological anomalies, as well as the gradients from moving water. The area of the body where this is sensed is in the vicinity of the sinuses and the ethmoid bone, which lies across the forehead above the eyes.

The scientific community in Europe has given far more scientific attention to the dowsing phenomenon, especially in France, where there are many active professional societies for dowsers and a museum in Paris displaying various dowsing pendulum instruments used throughout the centuries. More interest is being shown in the United States, too.

By Land and by Sea: Nature's Electromagnetic Sixth Sense

The more science looks at the electromagnetic influences in nature, the more astounding the connections become. It would appear that every living thing is tuned in to the earth's electromagnetic background and utilizes it for a variety of purposes.

The supersenses of birds and animals were once thought to be supernatural. Today many of these heightened senses are understood to be physiological, with electromagnetic properties high on the list. For instance, many animals can foretell earthquakes: snakes and scorpions seek shelter; cattle stampede; birds sing at the wrong time of day; and mother cats frantically move their kittens. It is now thought that they are reacting to changes in the earth's magnetic field, as well as to electrostatic charges in the air—long before the quake actually occurs or registers on even the most sensitive instruments.

The natural mineral magnetite, which is found in living tissue, seems to play an important role in explaining many electromagnetic mechanisms. Magnetite reacts a million times more strongly than most other magnetic materials, such as iron. Many creatures manufacture their own magnetite, and it is also found in bacteria and protozoa. The lines on the bodies of fish are long chains of magnetite crystals. The eye area of most birds is loaded with magnetite. Sea mollusks have teeth of magnetite—which has led to the discovery that it is not only the eternal motion of waves that wears away a rock ledge along a shoreline, but also the action of mollusks eating at it.

Magnetite is found in mammalian cells, too, contrary to what had been considered scientific dogma until recently. In humans, chains of magnetite crystals have been discovered in brain tissue. The ethmoid bone above the eyes and sinuses has a high concentration, and so does the blood-brain barrier. (The discovery of magnetite in these areas is significant because of the proximity of the ophthalmic nerves, which carry much information to the brain.)

Other physiological features seem to affect animals' ability to sense electromagnetic fields. Whiskers, for instance, may actually function as antennas and, when combined with different animal body shapes, may influence how electricity is conducted in some animals. Gray seals, even when blind, can easily locate prey through their whiskers, and insect antennae may literally be just that with relation to electromagnetic fields, not just sensing devices for air currents as previously thought.

Animals also react very differently to electromagnetic fields, which has implications for animal research and its applicability to humans. Most animals avoid high-intensity fields in any frequency. Cats appear to like low-level electromagnetic fields, whereas dogs do not. Why? Is there something about the feline shape that conducts energy more or less efficiently? Animal pathology has turned up other startling information, such as the finding that tumors in laboratory mice are a thousand times more magnetic than other body tissue.

Depending on latitude, the strength of the earth's natural DC magnetic field is about half a gauss (or 500 milligauss), and this pulsates generally at around 10 hertz, which is low on the spectrum. (We will later be speaking in terms of 1 to 3 milligauss, but this will be alternating current, which has different properties and bioeffects.) In addition, the average fluctuation within the daily twenty-four-hour cycle is less than 0.1 gauss. For years, scientists thought that these strengths were far too low to influence much of anything, but that opinion has changed radically in recent years. Wildlife has been found to rely on electromagnetic fields or to interact with these micropulsations in many ways, using magnetic information to give them their sense of place and as a compass to guide their every move.

Birds. Birds' navigation tools are the position of the sun and stars and the earth's magnetic ley lines, which always indicate the position of polar north and south. But on overcast days birds navigate by means of the earth's magnetic fields alone.

Many species, including raptors, geese, and even that cheery harbinger of spring in northern climes, the robin redbreast, use the earth's magnetic fields to guide their migration routes. Barnacle geese, for instance, migrate over a thousand miles every year, from Greenland to Scotland. House martins annually migrate from Britain to Africa. Annual raptor migrations stretch from Canada to Florida and back. Arctic terns breed around the northern ice cap in summer, then fly eleven thousand miles to Antarctica for that region's summer season. These migration routes have existed for millions of years; each generation of a bird species is born with a preset migratory genetic clock and compass.

Migratory birds rarely get lost, but they sometimes end up thousands of miles off course due to disruptions in the magnetic fields, either naturally caused by storms or artificially caused by man. Homing pigeons, for instance, have been unable to find their destinations in experiments in which contact lenses or small magnets were attached to their heads; both items would interfere with their ability to sense the earth's magnetic fields.

Fish and Mammals. Underwater, an invisible electromagnetic landscape maps the contours of the ocean's peaks and valleys, and many kinds of fish, if not all, use these magnetic markers to navigate their way around. Fish and sea mammals have also developed other electromagnetic senses that work in synergy with each other—not surprising, since their medium, water, is a natural conductor of electricity.

Dolphins and whales use both the earth's magnetic contours and the sound frequencies of sonar to navigate and to communicate. Those graceful batlike fish called stingrays have tiny pores all around the mouth and face that can detect the smallest magnetic field or electrical discharge.

All life generates electricity through nerve impulses and muscular movement, and water creatures have especially adapted to detect such minute neuronal electrical discharges in their pursuit of food. Stingrays have sensors that detect minute electrical discharges from crabs and crustaceans. And the platypus—that truly odd mammal, which lays eggs like a bird and has a bill like a duck—has sensitive electrical sensors in its nose that can detect an electrical discharge of less than one one-thousandth of a volt from a tiny shrimp several feet away.

Some lakes may support as many as two hundred different species of fish that use electromagnetics to find food. Many fish, in fact, have evolved to harness their own electricity. Specialized muscles in some fish can generate a 3-volt field, which is used to detect anything around it, in essence forming an instant electrical readout of an area. The long body of the electric eel generates a low-volt field for navigation, and the electric eel can step up its voltage to capture food. It can deliver a 500-volt zap to kill its prey.

Sharks have a whole array of electromagnetic tools. Their eyes are ten times more light sensitive than those of humans; they can detect subtle changes in the earth's magnetic field and can sense electrical neuronal discharges in living things. In fact, it is these electrical discharges that guide the shark's final attack. But sharks are often fooled by artificial signals. Untold numbers are killed annually when they attack underwater electric cables, mistaking them for live food giving off electrical impulses.

Schools of fish use electromagnetic senses to swim in synchrony. Each fish creates its own bow wave, which others in the school detect through what is called a lateral line, extending the length of their bodies. Lateral lines are like long biological bar magnets composed of chains of magnetite crystals. Lateral lines sense many things, including predators. Blind cave fish can "see" their surroundings through the magnetic sensing of their lateral lines.

Bats also use a combination of electromagnetic sensors in different frequencies. Like whales and dolphins, they navigate with the aid of sonar, which sends out sound waves that bounce off objects in their path and provide a constant in-flight readout. Vampire bats have special heat sensors in their noses to guide them to the richest, most easily accessible blood veins in their prey.

Insects, Amphibians, and Others. Like birds, many species of butterfly migrate thousands of miles annually; the best known is perhaps the monarch, which flies from the Hudson Bay region of Canada to the tropics of South America, crossing the Caribbean en route. Locusts also migrate thousands of miles. Such insects use their antennae to sense air currents and, most likely, the earth's electromagnetic fields.

Moths are drawn to the light frequencies. They position their bodies in flight to keep light on one side—the reason they fly around and around a streetlight, singeing themselves to death when they accidentally touch the bulb.

Among their many industrious talents, ants have been shown to be adept at electrical transmission. In experiments ants were found to organize themselves and align the antennae on their heads in an effort to channel and thereby mitigate electrical fields artificially imposed on them by researchers. The ant alignment behavior was seen at frequencies as low as 9 megahertz. (The field intensity was a 3-centimeter electromagnetic field, which did not surprise some researchers, since ant antennae are a kind of "quarter wave" resonator at 3 centimeters. These terms will become more familiar later on.)

In addition, flying ants—which are the breeding form of a colony—are incredibly sensitive to fluctuations in electromagnetic fields and atmospherics. These ants emerge for one day only and mate in the air, for which conditions have to be perfect. They are able to detect a calm, stable day from well below ground and then make their way to the surface.

Salamanders and turtles have incredible navigational abilities that are most likely based on magnetic sensing as well as smell. Both return to the exact spot where their own birth shells were hatched to mate and lay their eggs. For some species of California salamander, this is a yearly thirty-mile journey; for some species of turtle, it is thousands of miles. Experiments have shown that salamanders can even compensate for artificial interference by orienting themselves quickly and correctly to the earth's geomagnetic background.

Bees have been proved to have a fascinating sense of the earth's magnetic DC fields and the position of the sun. Karl von Frisch won the Nobel Prize in 1973 for his work on the honeybee "dance," in which he showed that bees orient themselves by using the angle of the sun and the center of the earth (a gravity vector) to perform a complex dance telling others in the hive where pollinating flowers are located. Bees were also found to have a detection system for polarized light to determine the sun's direction even through a forest canopy or on overcast days. Subsequent research located thick clusters of magnetite in the abdominal areas of bees, which function as built-in magnetic compasses.

In several experiments, honeybees have shown adverse reactions to electric, though not necessarily magnetic fields in hives positioned under high-tension lines. Hives in wooden casings were found to have an increased production of propolis, a resinous substance that bees gather from plants and use as a sealer. Colony sizes decreased, while irritability and mortality increased and over-winter colony survival was poor. The bees may have been experiencing low-level shocks, since the placing of grounded mesh wiring over the hives reduced many of these effects. But these particular researchers said that other interactions may have been occurring that were too subtle to measure. European researchers observed that bees subjected to certain electromagnetic fields were driven into a frenzy and attacked each other, as well as showing other adverse reactions.

Snails are another creature with magnetic sensors. They alternate between solar and lunar cycles, using their antennae to subtly influence their sense of direction.

In 1975, Richard Blakemore in America discovered that some bacteria also have a magnetic sense, which they use in seeking to dig deeper in the earth through mud. (His findings were later confirmed by research in Brazil and New Zealand.) Magnetotactic bacteria (those that react to magnetic fields) were found to contain straight lines of magnetite microcrystals surrounded by a thin membrane.

In any of the higher species in which magnetite crystals have been found—which is twenty-seven and counting, including three primates—one consistent observation has been that rich nerve fibers extend in many directions from the magnetite. This has led some scientists to speculate that multisystems rely on magnetic information and that their "need to know" is varied and complex.

With so much evidence from so many diverse species that magnetic sensors not only exist but are crucial to survival, the question of whether all species rely on such mechanisms arises. This is beginning to look more and more likely, but as with all significant discoveries, it raises more questions than it answers. For instance, how does the nervous system read the information in the crystal structures and then translate it into directions? And what aspect of the earth's natural magnetic information is being read?

If we better understood the answers to these and other questions, we would have a clearer notion of the perils that artificial EMF sources may pose for all the earth's creatures, humans included.

Chapter 5

TREES AND PLANTS:
THE THINGS THAT SUSTAIN US ALL

TREES, PLANTS, and the entire world of vegetation are also delicately entrained by electromagnetic fields, and some truly fascinating if not outright esoteric research has turned up thought-provoking results whose implications are staggering.

Every high-school biology student learns that plants use the light frequencies to produce their greenery through photosynthesis. But light is probably not described as only one of several frequencies on the electromagnetic spectrum at that point in the student's education—if it ever is. Yet the range of frequencies that plants utilize is much greater than that of light alone. In fact, only recently have some researchers come to believe that plants may have nervous systems that function not unlike those of human beings. A research group in Heidelberg, Germany, has discovered chemically unstable signaling materials called turgorines, which are apparently to plants what neurotransmitters are to humans. Turgorines regulate such plant functions as water distribution and leaf closing or drooping at sundown, a plant's equivalent of sleep. It also appears that, as with humans, a plant's nervous system can suffer damage from external sources.

All life-forms function according to their individual DNA blueprints. These blueprint codes are locked within the DNA molecules of a cell's nucleus. This is true of plants as well as both higher and lower life-forms. Life cannot be preserved if this DNA information is lost or has been interfered with to the extent that it cannot copy itself or be transmitted. DNA copying is not happenstance. In order for cells to do something, they need to be told first what that something is—not unlike a computer with hundreds of complex functions that sits idly purring away until an external stimulus (the computer operator) comes along to give it directions. The something that tells plants what to do is electrically based, as it is for all life-forms.

Electricity and Plants

The effect of electricity on plants has fascinated researchers since the eighteenth century. Around 1747, a French physicist named Jean Antoine Nollet theorized that plants were nature's hydraulic machines, after he observed that water uptake increased in plants he had placed in electrified pots. In a series of experiments, he also found that electrification influenced seed sprouting. Mustard seeds placed in electrified tin containers for one week sprouted and grew about 2.25 millimeters tall, whereas nonelectrified seeds sprouted and grew only slightly. Nollet theorized, in a presentation to the French Academy, that electricity profoundly influenced growth functions by somehow altering the viscosity, or flow resistance, of fluids within living organisms.

In 1770, in Turin, Italy, a Professor Gardini hung a series of wires over a monastery garden. Within a short time, many previously healthy and productive plants began to wither and die, but revived when the wires were removed. Gardini speculated that the wires either had interfered with some natural electric supply or had provided too much electrical stimulation from an external source. Around the same time, an Italian physicist named Giuseppe Toaldo reported that two jasmine bushes located next to a lightning conductor had grown thirty feet tall, whereas other jasmines in the same row were only four feet tall. (This was during the same period in which Galvani and Volta were making their ground-breaking discoveries.)

In the early 1820s, William Ross in America also found that seeds placed in an electrified medium sprouted much faster than nonelectrified ones. And in the 1840s, an agronomist named Edward Solly got mixed results in over seventy experiments with various grains, vegetables, and flowers, using both overhead and underground wires. In nineteen of the experiments, he got positive results but an almost equal number of adverse reactions.

Between 1868 and 1884, a Finnish scientist named Selim Lemström made a series of expeditions to polar regions. His observations there led him to theorize that the rich vegetation during the summer months was not due to the long daylight, as had been thought, but, rather, to the increased electrical activity of the aurora borealis. Lemstöm hypothesized that the sharp points of plant leaves functioned like lightning rods to facilitate the exchange of ions between the air and the ground. (We will shortly see that he was right.) To test his theory back at home, Lemström designed a system of flowers in metal pots, which he connected to overhead wires positioned sixteen inches above them and which in turn were grounded in the soil. After eight weeks, the electrified plants weighed nearly twice as much as the plants in a control group of pots outside the network. He then transferred the network to a garden plot and nearly doubled the normal yield of strawberries; the barley yield increased by one-third. He also reported that the strawberries were unusually sweet.

Lemström conducted many further experiments, which he published in 1902, in Berlin, in a book called *Electro Cultur* (an English edition two years later was titled *Electricity in Agriculture and Horticulture*). Even then, Lemström observed the difficulties presented by the scientific cross-nature of his subject, which spanned physics, botany, biology, and agronomy. The same problem still plagues the subject of electroagriculture (as well as electromagnetic fields in general), although today it has become mostly the purview of electrical engineers.

Others continued the experiments with electricity and crop yields. The English physicist Sir Oliver Lodge grew plants under a grounded-wire network that could be raised as the plants grew taller. He reported a 40 percent increase in wheat yield and, like Lemström, noted that the wheat was of superior quality. A collaborator of Lodge's named John Newman got an increase of 20 percent in wheat and potato yields. His work was eventually publicized in American engineering, rather than botanical, circles.

It had become obvious to these early researchers that the amount, kind, and duration of electrical stimulation were critical to the plant changes they observed. But their measuring devices were primitive by today's standards, making this early work very imprecise.

What had become known, however, by the end of the eighteenth century was the existence of vast atmospheric electricity and that the soil constantly emits electrically charged particles into the air.

Plants, EMFs, and Some Alarming New Findings

Research in the latter half of the twentieth century on the connection between plants and electromagnetism has likewise produced intriguing—and disturbing—results.

In the 1950s, a Hungarian refugee named Joseph Molitorisz became fascinated by some of Nollet's earlier theories about plants and electro-osmosis. In particular, Molitorisz wondered how the sap of the giant redwood tree could flow up more than three hundred feet, when even the best state-of-the-art hydraulic suction pumps could achieve a mere tenth of that. Molitorisz believed there must be something about electricity and trees that was turning the laws of hydrodynamics and engineering upside down.

Working at a U.S. government agriculture research station near Riverside, California, Molitorisz ran electrical current through citrus seedlings. The small trees increased their growth rate when the current ran in one direction, but, amazingly, they shriveled when the current ran in another. This important discovery signaled to Molitorisz that artificially applied electrical current could either bolster or block the natural electrical flow present in trees. He next designed an experiment with larger citrus trees in which he connected six orange branches to a 58-volt direct current but left another six branches nonelectrified. After eighteen hours, he found that sap flowed in the electrified branches but hardly at all in the untouched ones. Molitorisz wanted to apply this discovery toward the common problem for growers of having all their citrus fruit ripen at the same time. If growers could electrically stimulate some trees but not others, ripening could be staggered; green fruit would remain on some trees, while ripe fruit on others would fall to the ground. But he was unable to get additional funding.

Working on the other side of the country, in the Materials Research Laboratory at Pennsylvania State University, Dr. Larry Murr discovered, through first-ever laboratory simulation of brief thunderstorms and extended rainy weather, that growth in plants was directly tied to specific voltages. He also found that other voltages would damage plants. Furthermore, an electronics engineer named James Lee Scribner, in Greenville, South Carolina, produced a butter-bean stalk that grew twenty-two feet high and yielded two bushels of beans when he wired its aluminum pot to an ordinary electrical outlet. Scribner theorized that it is electrons that make photosynthesis possible, by magnetizing chlorophyll and thus enabling the plant to use solar energy. He also reasoned that magnetism draws oxygen molecules into the chlorophyll cells as they expand and that moisture uptake is purely an electrical phenomenon.

Magnetic forces have been found to influence plants in many ways. There are indications that plants align themselves with the earth's magnetic fields in the same way as higher life-forms do. In 1960, a professor of botany at London University's Bedford College discovered that plant roots are sensitive to magnetic fields, as described in a paper published in *Nature* called "Magnetotropism, a New Plant Growth Response." About that time, Russian researchers discovered that tomatoes ripened faster when placed closer to the south pole of a magnet than to the north. And in Canada, Dr. U. J. Pittman, at the agricultural research station in Lethbridge, Alberta, noted that many varieties of natural grains and weeds, all across the continent, consistently aligned their roots in a north-south direction. He also verified that many different kinds of seeds, when planted with the long axis pointing north, germinated significantly quicker—lending credence to all the old folk wisdom about how to lay seeds in the ground.

Magnetite apparently plays a role here, too. When infrared photographs from NASA satellites indicated that weakened wheat fields had a very different electromagnetic signature than healthy fields, a Colorado space engineer named Dr. H. Len Cox became intrigued by the notion that a natural magnetizable substance could be used to affect plant growth. Cox ground ferrous ore—magnetite—into a fine powder, added some trace minerals, magnetized the concoction, and applied it to a garden plot filled with root vegetables—radishes, carrots, turnips, and rutabagas. Although the tops of the plants remained normal, the underground changes were astounding. Not only were the vegetables on average twice the size of the control plants, but their taproots were nearly four times longer. The magnetite-activated soil also stimulated green plants like beans, broccoli, and lettuce into increased vitality. And, as with Lemström's earlier work, the produce was reported to be more flavorful.

For some years, the chemical mixture was sold to various users, who also reported astounding results. Then an interesting anomaly came to light. The mixture did not work when used in flowerpots or greenhouse flats, but had to be worked directly into the ground.

Electroagriculture in the Future

Even with so much promising electroagriculture research coming from different countries, the field continues to be a scientific curiosity, with no major sources of funding. The reasons are more than likely economic. Agribusiness as it is practiced today is chemically based, and its success is thought to depend on chemical fertilizers added to the soil and pesticides applied to plants to avert insect damage. Perhaps the new emphasis on less use of chemicals will stimulate more interest in electrobotany. But massive electrical grids strung for miles throughout the grain belt would not be feasible economically—or aesthetically. And such grids might well generate unsafe electromagnetic fields around them.

But smaller applications of direct current, to orchards, say, might be possible, and magnetite mixed into the soil surely would not contaminate the nation's water supply the way chemical fertilizers do. An added benefit is that healthy crops resist insects better, reducing the need for pesticides.

Much remains to be learned before farmers begin to electrify their fields willy-nilly. As has been consistently true with other species, numerous researchers have found that EMFs have a variety of effects on plants and crops, including detrimental ones. What is needed, of course, is a bridge between the sciences of physics, electrical engineering, botany, and agronomy to bring under one roof the work now going on in these divergent disciplines.

Corona Effect, Tree Damage, and Deforestation

As we have seen, the effect of electromagnetic fields on plant life is a two-edged sword. EMFs can have both a positive and a negative influence on plants, just as they do on humans, and it is conceivable that they are causing irreparable damage by changing the ecosystem of the world as we know it. For instance, electromagnetic fields are known to adversely affect the pheromone levels of trees and plants, making them more susceptible to insect damage. But there are indications that we have already gone beyond this rectifiable stage and into far more serious territory.

In the mid-1970s, research was initiated specifically to investigate the effects of 60-hertz fields on plants. Sharp-pointed plants were found to act as lightning rods, just as Lemström had speculated around the turn of the century. A research team from Westinghouse Electric Corporation and Pennsylvania State University studied eighty-five laboratory-grown plant species exposed to fields of 30 and 50 kilovolts. Damage to plant tips was observed at strengths of 15 to 20 kilovolts. The strength of the field was greatly enhanced around sharp-pointed leaves, which act as conducting objects by causing the field lines of force to be drawn to and converge around the sharp points. Rounded plant parts do not produce the same effect. (Grass is susceptible because its leaves are single strand and pointed—explaining why grass often doesn't grow under certain types of electric fences.)

If the electric field is strong enough, something called a corona effect will occur. Corona is a sort of electrical leakage or discharge, which can be exacerbated by the presence of water. Plants, of course, have a high water content, especially in their new-growth areas, like the leaf tips. When plants with pointed tips (already making them natural electric conductors) are subjected to higher electrical fields, like those around high-tension lines, the tips can suffer from an induced corona effect.

Experiments with pine trees allowed to grow unusually high within high-tension line rights-of-way have shown branch damage from induced corona. The needle tips dry out and die back in a kind of self-pruning, as if to escape from the electric field, while the rest of the tree appears to survive. Such studies have been largely confined to the 60-hertz frequency range, but far more serious and perhaps permanent damage may be occurring from the microwave/radio-frequency ranges.

While a certain amount of the sun's natural rays (cosmic "white noise") that hit the earth falls within the microwave/radio-frequency ranges (and therefore presumably has an influence on global plant life), the artificially induced frequencies that bombard our flora and fauna have very different properties that do not exist in nature.

The higher frequencies used for radio, TV, radar, and cellular-phone transmission are of special concern. The transmission intensities are sometimes on the order of a billion times more powerful than what exists in nature; the frequencies transmitted are extremely specific; the waves are polarized; and the signals can be digitally and analogously modulated. (Digital technology functions in a kind of on/off way that simulates a pulsed wave, which has been found to be more detrimental to living systems.) These special features can theoretically interfere with any number of biological information systems by acting as a kind of scrambler.

Dr. Wolfgang Volkrodt, a retired Siemens physicist and engineer in Germany, has led the environmental dialogue regarding EMF damage to trees over large areas of deforestation in the German and Swiss Alps. Dr. Volkrodt's theory holds that this extensive deforestation, which was previously attributed to acid rain, has another underlying cause, since the same kind of damage exists where no acid rain falls. Dr. Volkrodt believes that an electrolysis process occurs in forest soil, due to the absorption of electrical current by pine trees through their needles and by deciduous trees through their leaf veins.

Electromagnetic fields are "received" by trees. While not perfect electrical conductors, trees are capable of channeling electric fields through their needles or leaves, into their trunks, and down into the soil through their roots. Dr. Volkrodt believes a resonance relationship is present.

When resonance (the phenomenon that can occur when an aspect of a force, such as a frequency wave, matches some characteristic of a physical object) occurs, the power inherent in the force or wavelength is maximally transferred to the physical object, causing it to resonate or vibrate. Within the object, the resonance is self-perpetuating. An opera singer hitting high C and causing a crystal goblet to shatter is a dramatic illustration of the resonance phenomenon. But objects can resonate without shattering.

Dr. Volkrodt has determined that the shorter the wavelength, the better it matches the physical structure of a tree and the more damaging it is. The microwave band corresponds almost perfectly to the resonance capacity of the cell membranes of trees. Such resonance could interrupt water circulation, making the tree incapable of raising water or sap. Trees die from the top down; they can literally be standing in water and die of thirst.

Dr. Volkrodt sees the possibility of permanent damage to the soil when the millions of organisms that make up a healthy forest soil are affected by this electrolysis process and undergo a change in ion balance. In effect, the soil becomes dead, and new growth is hindered. He believes that this effect will continue until the electromagnetic contamination stops. Unfortunately, the major communications companies' vision of a wireless future rests completely on frequencies that fall within that EMF range.

The kind of forest damage that exists in Germany may be a by-product of the heavy concentration there of radar and the myriad listening technologies of the cold war. But such deforestation is present in Canada and the United States, too. Sometimes it is a by-product of commercial communications companies' bouncing their signals off mountain ridge lines—a common practice. Other damaged areas can be found around military installations. But no one in the United States has focused on EMFs and environmental damage as Dr. Volkrodt has—perhaps because so few of his papers have been translated from the German.

If his theory proves accurate, the implications are dire. The world's forests are the flip side of human life. We produce carbon dioxide, which plant life requires; plants synthesize it and give us back oxygen, without which we cannot live. In order for this vital cycle to continue, it is estimated that each human being requires approximately four full-grown healthy trees capable of photosynthesis. With the world's human population increasing at a staggering rate, our need for oxygen will only increase. Clearly we cannot afford to have deforestation continue, especially if the cause is one we can control.

Dr. Volkrodt recommends a simple solution: cease wireless communications and use fiber-optic cable instead.

PART TWO

THE
MEDICAL
ISSUES

Chapter 6

WHAT YOUR DOCTOR
DOESN'T KNOW, AND WHY

T HE OLD ADAGE that the more things change, the more they
remain the same is particularly apropos regarding our evolving understanding of electromagnetic fields and the human
body. The cutting edge of Western medical research today is returning to realms that are five thousand years old.

Many contend that today's mainstream medicine has a certain lifeless sterility to it, that it has been taken over by techno-
crats and their technology, trained to think mechanistically and
to compartmentalize the human body in a way that has little relationship to the whole of a person's psychophysical experience. We
seem to know more and more about the machinery of the body
and less about the life within it. We act as if life's mysteries will
reveal themselves if we just pay enough attention to physical mi-
nutiae, as if the presence of life is an automatic offshoot of the
body's good mechanical design and the absence of life an unfortu-
nate design flaw.

The most thorough study of anatomy will not reveal the se-
crets that cause life to animate each of us. So compartmentalized
and illness oriented has mainstream medicine become that the
very concept of a diseased body acting as a whole is upsetting to
physicians and patients alike. But life is far more than the de facto

absence of death, and for all our advances, we actually know very little about it.

To some extent, mainstream medicine is a victim of its own success. Its overspecialization stems from the great advances in knowledge about particular aspects of the body. But such specialization has led to a nonintegrated system that often functions in a practical as well as a visionary vacuum. Medical practitioners rarely read outside their own specialties (who has the time?), and so advances in one area that might have a profound benefit for another can remain undiscovered for decades. This is compounded by the lag time between actual laboratory research and publication of the findings, which is typically two years at a minimum—and another five to ten years before a consensus on treatment is reached.

Record numbers of Americans today are reacting to the aridness of mainstream medicine by turning to some form of alternative medicine. They are seeking out naturopaths, homeopaths, and Eastern practitioners, whose treatments often prove effective for certain chronic disorders that fail to respond to mainstream medicine. (An even higher percentage of Europeans, including members of the British royal family, uses such medical alternatives.) So popular have these modalities become in the United States that the National Institutes of Health has formed a special division to investigate their legitimacy.

What Primitive People Knew:
Vitalism and the Beginnings of Western Medicine

Today's mechanistic medical model lies at the end of a long road that begins even before recorded language. What we have lost along that road and are currently rediscovering is primitive people's fundamental awe for life itself.

Most primitive cultures believed that a special energy animated all living things—humans, animals, and plants—which in turn possessed and embodied this primal force.

Certain gifted individuals, such as shamans, priests, and Yi Qi Gong masters (in the East), were thought to be able to channel this life force for the benefit of others. And it was believed present in a far-reaching pharmacopoeia of herbal remedies as well as in some nonliving objects, like the lodestone (a naturally magnetic rock containing magnetite).

The primitive world found evidence of this unseen force in such natural events as solar and lunar changes; birth, aging, and death; tornadoes, earthquakes, volcanoes, and floods; and the changing of the seasons. All manner of beliefs, superstitions, and rituals developed in an attempt to explain man's place in it all. These early explanations would later divide into philosophy, religion, and medicine.

Natural forces were understood to have an effect on humans. Intermediaries (like shamans) armed with powerful tools (like the lodestone) could mediate between this natural universe and the individual. What has been lost between primitive cultures and our contemporary one is the vitality that drove the early quest to understand humankind in the cosmic scheme of things. Some of this vitalism is now being restored—though not in its primitive and superstitious forms—in the growing understanding that electrical and magnetic forces are as central to the human anatomy as they are to the cosmos around us. We are scientifically relearning what the ancients understood on an intuitive level.

Ancient Texts and
the Origins of Energy Medicine

The oldest known medical text is the Chinese *Yellow Emperor's Book of Internal Medicine,* attributed to Houang-Ti and dating from around 2000 B.C. It describes a sophisticated system employing techniques that had a therapeutic impact on *chi*—a complex bodily energy that worked through the balance of the opposing universal forces of *yin* and *yang*. An imbalance in these two forces resulted in disease, which very detailed modalities could cure by restoring the balance. These were acupuncture, the insertion of fine metal needles into specific energy points along well-defined lines called meridians, accompanied by the placing of lodestones on the energy points; and moxibustion, the burning of small amounts of herbs over the same energy points to produce sweating. According to Dr. Robert Becker, author of *Cross Currents: The Perils of Electropollution, the Promise of Electromedicine*, these techniques were believed to influence the body's internal energy system by introducing external energy to treat disease states. The forms of external energy were electric (the metal acupuncture needles), magnetic (the lodestone), and thermal (the heat generated from the burning moxa). Dr. Becker speculates that the origin of these treatments preceded the *Yellow Emperor's Book* by several thousand years, attesting to the sophistication of preliterate peoples.

Other cultures also made use of electromagnetic healing methods. Cleopatra is said to have worn a lodestone on her forehead to prevent aging, and Egyptians used lodestones in various therapies dating back to 2000 B.C. Around the same time, the Hindu religious scripts called the Vedas describe the treatment of illness with *siktavati* and *ashmana*, which can be translated as "instruments of stone" and may well have been lodestones. In addition, Tibetan monks use bar magnets in training the minds of noviates, a practice that likely goes back thousands of years and probably originated with lodestones. The Tibetans also teach an esoteric meditation technique done in prolonged darkness, which undoubtedly stimulates the brain's light-and-magnetic-field-sensitive pineal gland.

Many of the ancient healing techniques that made use of energy methods and of electricity and magnetism are still in wide use today in Asia.

The Genesis of Western Medicine

Western medicine traces its origins to ancient Greece, to the work of Thales of Miletus (ca. 624–546 B.C.). Considered the "father of philosophy," Thales laid the scientific foundations for what would become physics and biology, besides making several discoveries in geometry and astronomy. He was the first to write about the lodestone and static electricity, which he discovered by rubbing amber with fur (the Greek for "amber" is *elektron*). To Thales, the magnetic potential of the lodestone and its ability to attract iron indicated the presence of a soul of sorts, and he thought all living things were animated by a similar vital spirit. He also searched for a primary substance in the universe that would explain both change and stability and believed that water was this substance. But Thales' great contribution to Western science was in his approach to inquiry, in his introduction of the idea of causal relationships arrived at through reason, logic, and observation rather than belief and superstition.

Hippocrates (460–377 B.C.), the great physician often called the "father of medicine," left a prolific legacy of eighty-seven treatises and a code of medical ethics. (The version of the solemn Hippocratic oath sworn to by all medical-school graduates today still enjoins them to "First, do no harm.") Hippocrates, who drew much from Thales' legacy, could almost serve as a quintessential model of the family doctor. He was knowledgeable but not arrogant, ethical and curious about his patients but humble in the face of what he did not know. To him, a disease state was a complex relationship between an external agent and the body's reaction to it, a fluctuating condition interacting with the patient's self-healing proclivities. Life was a vital spirit that acted through the four humors of blood, phlegm, yellow bile, and black bile; an imbalance between these humors led to disease (not unlike the Eastern notion of imbalance in *yin* and *yang*, as well as certain other Asian practices such as Indian Ayurvedic medicine, which preceded Hippocrates by several centuries). He used herbs in treatments, probably lodestones, and perhaps even something similar to moxibustion (based on some of his writings concerning iron and fire in treatment techniques).

Another of Hippocrates' achievements was the founding of the first medical schools, called *aesculapiae,* throughout the eastern Mediterranean. In the third century B.C., one of these schools, in Alexandria, Egypt, produced an exceptional physician named Erasistratus, who is considered the first scientist to dissect and accurately understand the human anatomy.

Erasistratus was responsible for correctly identifying motor and sensory nerves, the function of the brain to which they were connected, and the heart as a pump for blood.

Whereas Hippocrates had thought that the heart was the seat of the mind and soul, Erasistratus attributed them to the brain. He was a vitalist who believed that a subtle vapor, which he named *pneuma,* was responsible for the life force, but his mapping of the mechanics of the body led him to attribute disease states to abnormalities of various organs, rather than to Hippocrates' imbalance of the humors.

Erasistratus' views prevailed for a few hundred years, but they eventually were usurped by the arrogant, self-righteous, self-serving Galen, a Greek of the second century A.D. The theory for which Galen is perhaps best known is that for every disease there is a single cause and a single cure—the first step toward the medical authoritarianism that prevails today.

Galen's legacy included a complete system of medicine, incorporating anatomy, physiology, and treatment, that became standard dogma for the next 1,500 years. Galen attacked Erasistratus' work directly, repeating and falsifying some of Erasistratus' painstakingly careful experiments. But so prominent had Galen become that no one dared repeat *his* experiments. Much of Galen's writings have been preserved while the body of Erasistratus' work has largely been destroyed.

What was lost to science under the authoritarianism of Galen was the humanism of Hippocrates and the logical observations of

Erasistratus. In Dr. Robert Becker's words, "the first wrong turn had been taken." With a comprehensive but erroneous system of medicine given an official stamp of approval, Western medicine entered the Dark Ages.

Scientific Medicine Emerges

Medicine's Renaissance, its emergence from the Dark Ages, is embodied in the career of a visionary radical named Paracelsus (1493–1541), a Swiss alchemist and physician said to be the model for Dr. Faustus, the legendary character who sold his soul to the devil in exchange for knowledge.

Reported to have left home at the age of fourteen, he wandered across Europe and Asia, studying at many universities. He hated authority of any kind but Paracelsus especially despised Galen and is said to have once burned his books in front of a cheering crowd of medical students. Paracelsus held fast to the notion that Galen's teachings could lead to disastrous complications for a patient, and that self-healing was a fact to be enlisted by any physician. He also accurately illustrated that mercury cured syphilis, thereby foreshadowing the use of antibiotics, and he was responsible for correctly describing the cause of thyroid goiter. He also laid the early foundation for homeopathy with the belief that small amounts of chemicals that produced symptoms similar to the disease would provide a like cure. In addition, he foreshadowed environmental/occupational medicine, by saying that silicosis in miners was caused by the mine's materials rather than angry mountain spirits. Some early foundations in chemistry were also formulated by Paracelsus through his experiments in alchemy and herbal remedies. Plus he used lodestones frequently in his practice.

It took strong will to weather the attacks of established medicine at the time. His popularity was such that his lectures, which were open to the public, consistently drew large crowds. His writings were both prolific and influential, the best-known of which is *The Great Surgery Book.*

Paracelsus wandered into some mystical territories that seemed unprecedented during that period in the West, indicating he may have studied esoteric Asian philosophies. He was a vitalist who passionately believed in a life force that he called *archaeus,* which he thought could be influenced by the action of magnets — long before the discovery of magnetic fields and their influence on every living thing. He also apparently believed that one's thoughts could have a physical effect, foreshadowing some current research in psychoneuroimmunology and parapsychology.

Unlike Galen, who Paracelsus considered a reductionist fool, Paracelsus saw the body as a whole system of many interrelated parts, a system inseparable from the total organism but greater than the sum of its individual components. He was the first holistic practitioner, enlisting herbalism, logical observation, and experimentation, as well as natural magnetic forces, into his therapies. His writings, along with his strong visionary direction, set the groundwork for much of our scientific inquiry today. Yet he died practically penniless and under mysterious circumstances at the age of forty-eight, leaving his few possessions to the poor and his manuscripts to an ordinary barber-surgeon.

The Age of Science and Reason

Within a few years of Paracelsus' death, Galen's infallibility was finally put to rest with the advent of the first truly accurate anatomical text, *De humani corpus fabrica (The Fabric of the Human Body),* by a Belgian surgeon named Andreas Vesalius. Gradually the sciences of living organisms (biology) and of nonliving things (physics) emerged as separate fields of inquiry, and the hitherto unfathomable forces of electricity and magnetism, once steeped in superstition, began to be understood as natural forces.

Some of the great names in science during this time are known to us still and their legacy forms the underpinnings of today's medicine and physics.

William Gilbert, who was Queen Elizabeth I's physician, was the first to accurately combine medicine with electricity and magnetism. In 1600, he published *De Magnet* (*The Magnet*), which clearly defined electricity and magnetism as separate forces, and the basic laws governing each. He also identified the earth itself as a giant magnet.

Several methods for storing static electricity, the only kind known then, were devised during the 1600s, and basic biological understandings about the nerve's transmission of sensory information resulting in muscle contraction, also became known. Finally, the brain was resolutely identified as the center of memory and thought processes.

Before long, the divisions between the mechanists (those who believed that living organisms were complex machines that could be totally understood through physical laws) and the vitalists (those who believed that there was a mysterious life force involved with living organisms) became more vehement, although the lines were not yet completely drawn. For example, René Descartes (1596–1650), the great French philosopher and mathematician, was a prominent proponent of the mechanistic model but nevertheless put forth the notion that the soul was located in the pineal gland at the center of the brain. It is curious that Descartes located the soul there, for this is also the location of the Eastern concept of the third eye.

Descartes is best known as the originator of analytic geometry and for his dualistic division of the universe into mutually exclusive but interacting domains of mind and matter. His famous axiom, "I think, therefore I exist," was for him irrefutable proof of the existence of the mind (or spirit). The mind was subject to reason; matter was subject to mechanical law. This Cartesian dualism clearly dominates the Western philosophy of medicine today—although our scientists tend to leave out the mind and spirit altogether and rarely venture into "metaphysical" territory other than to disprove it.

Asian philosophies, which never embraced the universe in mind-and-matter dualities, have had no difficulty incorporating energetic systems into their traditional medical treatments.

The Death of Vitalism, Understanding Electricity, and the Fragmentation of Medicine

In the ensuing fervor of scientific discovery during the Age of Science and Reason, theories of vitalism were squeezed into ever narrower corridors for lack of physical verification. This would eventually prove its undoing since universal energy was an ephemeral substance best explained by an absence of physical proof. Nevertheless, a few energy theories pertaining to the human anatomy continued to be explored.

The Austrian physician Franz Anton Mesmer (1734–1815), like the Asians, theorized that all living things generated universal forces that could be transmitted to others through what he termed animal magnetism. He believed that it was possible to concentrate this magnetic force in certain people and use it therapeutically—and did so with such success that the medical establishment accused him of practicing magic. In 1784, King Louis XVI of France appointed an investigatory commission, which determined that Mesmer's power was due to simple suggestion. Mesmer's work was similar to that of Paracelsus', and in today's Chinese practice of Yi Chi Gong or the West's Therapeutic Touch, Polarity, or Reiki therapies, Mesmer's legacy would be better understood for the magnetic healing modality that it probably was. Unfortunately, he is almost exclusively remembered as the originator of hypnosis (also called mesmerism).

A contemporary of Mesmer's was Samuel Hahnemann

(1755–1843), a German physician, who is considered the founder of homeopathy but whose work actually built on Paracelsus' theories of "like cures like." Hahnemann developed a complex system of medicine based on "the law of similars," which is enjoying a renewed interest today and is popular in Europe even in some mainstream medical practices. Homeopathy, which administers minute quantities of substances to produce symptoms like those a patient is already suffering from, is thought to stimulate the immune system into action, thereby taking advantage of the body's own healing capacity. Hahnemann thought these infinitesimal amounts of substances reacted on an energetic level with the body's vital spirit, in much the same way that the lodestone was thought to work. (In fact, he was a proponent of the lodestone's use in treatment.) Hahnemann's theories have recently been re-explored and expanded on in a book called *Vibrational Medicine: New Choices for Healing Ourselves* by Dr. Richard Gerber.

The general debate between the vitalists and the mechanists continued into the 1800s. The vitalists thought that electricity was the scientific basis to verify the life force. They may have been partly right, but in adhering to this single notion, a strategic mistake was made, since if electricity were to be excluded, they would have nowhere else to turn. And that is just what happened.

The shy genius Luigi Galvani (1737–1798), a physiologist and anatomy professor at the University of Bologna in Italy, had studied static electricity (still the only kind known) for twenty years and was convinced that in it he had proof of the life force's being electrical in nature. Observing that frogs' legs contracted when they were connected to the spinal cord by metal wires, Galvani proposed that they were drawing electricity hidden within the nerves themselves, which he termed animal magnetism. What Galvani missed was that the muscle contractions occurred only when the wires were made of different metals. What he had actually discovered was direct current, but he didn't know it. This discovery has shaped our entire modern world ever since.

(Galvani also discovered a natural "current of injury," the process by which an injured limb will produce a negative charge at the injury site, which will later turn to a positive charge at the same site. The implications of this will be discussed later in the groundbreaking bone-regeneration work of Dr. Robert Becker and his colleagues in our own time.)

Galvani unfortunately put himself squarely on the line in announcing to the Bologna Academy of Sciences, in 1791, that the body's vital spirit was electricity flowing through the nerves. This unwittingly gave the mechanists an objective theory to attack.

Within two years, a physicist and colleague of Galvani's named Alessandro Volta, from the University of Padua, proved that what Galvani had discovered was a new kind of electricity, in the form of a steady current rather than simple static sparks. In Galvani's original observations, the frogs' legs had seemed to contract when the wind blew them against an iron railing. Volta demonstrated that Galvani had generated direct current between two different metals (iron and copper) and that the frogs' legs were mere junctions between them; as such they were dispensable. Since the frogs' legs were mainly composed of salt water, they formed an electrolyte conducting medium, and were only electrical channels between the two wires.

Volta disproved Galvani's animal-magnetism theory, and the noncombative Galvani was crushed. Volta himself, by his grouping of different metals, had discovered the storage medium known today as the battery. But Galvani *had* actually demonstrated an animal magnetism of sorts—frogs' legs can be made to twitch with no metal in a circuit just by bringing the muscle into contact with the cut end of the spinal cord. This current of injury is found in any injured tissue and is the beginning signal for all healing processes.

Among Galvani's many other accomplishments were the demonstration that electricity can be transmitted across space (rediscovered a hundred years later by Heinrich Hertz) and the first use of antenna wires to search for atmospheric electricity.

But like many other pure scientists, he was far ahead of his time and paid a high personal price for it. Like Paracelsus before him, Galvani died penniless and dispirited, all his property confiscated by the French, while Volta, supported under Napoleon, grew famous for his storage batteries.

Both men's names have, however, come down to us in such terms as "galvanic skin response" (upon which lie detectors are based), "galvanized metal," "galvanize into action," and "volts" and "voltage."

Galvani's work was verified thirty years after his death by Carlo Matteucci, a physics professor at Pisa. In painstaking experiments over a 35-year period, Matteucci proved that the current of injury was accurate, but he located it exclusively at the injury site, rather than in the nervous system, the way Galvani had.

Then, in the 1830s, a physiology student in Berlin named Emil DuBois-Reymond discovered that nerve impulses could be measured electrically. He believed that this directly identified the nervous system's activity with electricity, but soon thereafter a researcher named Herman von Helmholtz measured the actual speed of the nerve impulse and discovered it to be much slower and weaker than that conducted in a wire. What this meant was that nerve impulses could be measured electrically, but the impulses were not necessarily the passage of actual blocks of electrical particles.

In 1871, Julius Bernstein stepped into the picture with a different chemical explanation. He thought that the nerve impulse was not an electrical current at all, but, rather, an action potential. What was being measured as a nerve impulse was a disturbance in the ions (charged atoms of sodium, potassium, or chloride) of a cell's membrane, and this disturbance was what traveled along the nerve fiber. Bernstein thought that there was a difference between the inside of a nerve cell and the tissue fluid that surrounded it and that this difference created an electrical charge or polarization of the membrane cell.

Bernstein's hypothesis was shown to be essentially correct for all the cells of the body, but instead of becoming a solid building block of anatomical understanding, it unfortunately hardened into physiological dogma. It has since been assumed that this kind of electrical activity is the only kind present in the human body. It is still the prevailing view that direct electrical current cannot exist within the body and that externally generated currents cannot have any biological effect below levels that cause shock or heat.

However, recent work by Dr. Björn Nordenström, the former head of diagnostic radiology at Karolinska Institute in Stockholm, Sweden, has challenged this narrow view by his findings of an intrinsic electromagnetic system within the body similar to the meridian concept. And Dr. Robert Becker's theories on the body's inherent DC system will be discussed later.

By the end of the 1800s, microscopes had made it possible to observe the actual space between nerve fiber and muscle—a space eventually called the synaptic gap. The vitalists, still clinging to electricity as the explanation for the life force, reached into this tiny area of physiology to hypothesize that the passage of nerve impulses across the gap was electrical. Then, in 1921, physiologist Otto Lowei proved that this impulse passage across the synaptic gap was chemical, too—a discovery for which he received the Nobel Prize in 1936.

Vitalism had nowhere else to turn and was eradicated from the mainstream medical dialogue.

Dr. Becker, a student of Otto Lowei's, tells a revealing anecdote about Lowei in his *Cross Currents*. Lowei used to caution his students that some mysteries remained beyond scientific explanation, and it turned out that he had resolved some plaguing problems with his Nobel Prize–winning experiment in dreams over two successive nights.

Clearly we need some new models to help explain what we can observe about the relationship between electromagnetic fields and the human anatomy. Just because Bernstein was correct about the fundamental action of the nervous system and because Lowei was able to demonstrate a chemical basis for nerve action, it does not automatically follow that this is the only way nerves transmit information or that membrane polarization is the sole means by which electricity works within the body. And the problem-solving dreams of Otto Lowei need to be kept in the reality picture, too.

We have no medical model for the mechanism by which low-level electromagnetic fields can cause damage to the human body, yet such associational damage continues to appear. That should have red flags popping up all over the playing field. As a general rule of thumb, when what we *think* we know continues to fly in the face of what we experience, we need to go back to the drawing board and create a medical/scientific vision expansive enough to encompass the totality.

The Twentieth Century

By the beginning of this century, scientific medicine was firmly based on the chemical-mechanistic model. In 1909, Paul Ehrlich coined the term "magic bullet" for his discovery of an arsenic compound that cured syphilis, and he correctly predicted that lots of magic bullets for specific illnesses would follow. (This approach is still the primary one being pursued and funded by mainstream medicine.) The dramatic discovery of antibiotics only encouraged this as the sole path in treating illness. Simple, predictable, infallible cures are as inviting today as they were in Galen's age. Life has been reduced to the interaction of chemical mechanisms integrated by the central nervous system. Electricity and magnetism have been all but factored out of the medical model, much to our peril. As Dr. Becker would say, "The second wrong turn had been taken."

Electricity and magnetism quickly became the province of physicists and engineers. By the 1920s, technological miracles like electric lights and radio were sweeping the industrialized world. We understood how these powerful forces worked, and we applied them to all manner of devices. We thought that the only effect electricity had on the body was through strengths high enough to cause burns or shock. And we presumed that magnetic force could only influence other magnetic materials; since there was no known magnetic material in the human body, no damage could occur.

Dr. Joseph Kirchvink, however, working at the University of California and building on discoveries made in 1961, recently found magnetite in human brain tissue. This has introduced a whole new dimension into our model for the human anatomy's interaction with electromagnetic fields, both natural and artificially produced. Add this to the fact that iron and copper, both magnetic materials, are an integral part of human tissue and are necessary for health, and more red flags appear on the playing field.

In allowing electromagnetic technologies to become the unchecked driving force for modern society at the same time that medical science had completely disregarded electromagnetic forces as an influence on the life process, we have made what Dr. Becker would call "the third wrong turn." This last misdirection is what has brought us to where we are today. It has created a monumental gap between the medical professions, the physics community, manufacturers of EMF technologies, and government regulatory agencies. Medicine itself has in many ways become more technical and less scientifically curious.

Dizzying paradoxes have been created in this gap. Hospitals regularly lease roof space to radio and cellular-phone companies, but no doctor in the building raises health questions or is asked to sit in on the decision-making process. People in their mid-twenties and early thirties will seek help from ophthalmologists for extremely dry eyes (a symptom usually found in the elderly), but the doctor does not ask if the person works with a computer or in an electromagnetically charged environment, even though electrostatic fields are known to be drying agents. Pregnant women asking about the safety of VDTs are routinely told that "nothing is proven." Those who work under fluorescent lights all day complain of vision and concentration problems, but few internists know that fluorescent lights function through ballasts with high magnetic fields and that the ethmoid area around the nasal sinuses is loaded with magnetite and therefore is theoretically magnetic. Why should they? Such ubiquitous technologies have been developed with complete presumptions of medical safety.

But doctors themselves are not immune. Several cases of brain tumors have occurred in physicians who frequently used cellular phones.

Hippocrates long ago admonished new physicians to "listen to your patients . . . they are giving you the diagnosis." Yet, in a final paradox, patients who tell their doctors they think they know what has made them ill—a computer, a work environment, a chronic EMF exposure of some kind, a diagnostic procedure— are routinely told that it's impossible, it's all in their minds.

Except for a handful of tenacious researchers, mainstream medicine has largely abandoned any serious quest to understand the biological effects of electromagnetism. But this is slowly changing. What most doctors know about electricity today is that it can cause burns and electrocution but not necessarily that it is part of the body's most fundamental biological processes; that these processes below the body's "threshold noise level" rest on more subtle aspects of physiology than the chemical model or the mechanists ever envisioned, or that minimal external EMF exposure can have profound effects.

Many of the problems that crop up today are largely treated as engineering, economic, or legal difficulties. In truth, they are medical problems, and we need to understand them as such. When we do, those other concerns will fall into proper perspective.

Chapter 7

THE BODY'S ELECTRICAL SYSTEM

MOST PEOPLE don't realize that electromagnetism is inherent in the human anatomy. Every time a muscle moves, there are electrical neuronal discharges; brain waves are electrical (they are what an electroencephalogram, or EEG, measures); all the sensory information that moves through the body to the brain is electrical; those chemical messengers called enzymes, which keep the whole organism of the body informed about various aspects of itself, are electrically influenced; cell division is electrical; the signal for the body to initiate wound healing and bone growth is electrical; and the heartbeat is electrical. (Cardiologists have even discovered a particular electrical misfiring within the heart muscles just prior to a heart attack so distinctive that it is being used to predict subsequent heart attacks.)

In fact, as we have seen, all chemical changes are electrically based because they involve the transfer, sharing, or alteration of electrons at the molecular level. One could go so far as to say that there isn't anything that occurs in the human anatomy that *isn't* involved with electromagnetism in one way or another.

That there are biological effects from electromagnetic fields, no one in any scientific discipline disputes. The only debate centers around whether there are long-range adverse health effects, and if so, what are they and can they be reversed?

A Quick Overview

Thermal effects (the heating of body tissue) from electromagnetic fields have been known since the advent of the technology in the 1930s. Most of the controversy has centered around the existence of nonthermal effects, meaning bioeffects by some other means. Although some still deny the possibility of nonthermal effects, such naysayers are becoming less common in scientific circles. Most of the cutting-edge research today is in nonthermal areas and focuses on the magnetic component. This was not the case a few short years ago, when research almost exclusively concentrated on the electric component, with an eye on thermal effects.

It is important to distinguish between the two because thermal-effects-only are considered reversible and are relatively easy to measure, making it possible, therefore, to set exposure limits accordingly. This is not to deny that thermal effects are complex and may set off a cascade of reactions in the body. It is only to say that they are more easily measured and dealt with.

Nonthermal effects are another case altogether. Not understanding them means we do not understand something fundamental about the human anatomy, while we are constantly being exposed to increasing levels of EMFs. It is like holding the tail of a tiger and calling it an innocuous powder puff, but ignoring the animal it is attached to or what it's capable of doing.

Electromagnetic fields can affect many different body systems, and the relationships are complicated. Researchers are finding statistically significant increases in leukemia (about thirty studies), brain cancers (about twenty studies), melanoma (about ten studies), and male breast cancer (about five studies). The increases on average are small—between 1 and 5 percent—but whenever a consistent pattern above the expected appears, it is generally considered important, and often the tip of the iceberg.

It's important to point out that far more studies exist showing

EMFs' adverse results than do for the relationship between cancers and smoking, yet no one hesitates to make health recommendations against smoking. What is not understood, although there are some interesting hypotheses, is the mechanism by which physical damage could occur at low EMF frequencies and low power densities. According to what we've presumed about the human anatomy, this should not be occurring. Yet often when a researcher tests some aspect of the subject, the findings are worrisome. This holds true for research in a laboratory focusing on cells in petri dishes (which doesn't automatically translate to how whole systems function), as well as epidemiological surveys of a large group.

No one understands the mechanisms that initiate the wild cell growth of any cancer, although certain substances (like tobacco and asbestos) are known to increase the risk. In truth, we don't understand a lot about what causes a healthy organism to slide into illness, or what keeps another well.

Perhaps EMFs will be found to cause disease, to be a co-factor in promoting latent disease, or to be a co-promoter in combination with other carcinogens. But that is only one side of the coin. Electromagnetism is also being used in a variety of treatment modalities that show great promise. No one knows why some fields cause harm or why some appear to heal. The point is that the entire subject should be approached with caution until we know more.

Physical Mechanisms

One of the areas of contention between consumer advocates, who want more government protection from EMFs, and the scientific community, upon which such government protection would rest, is *how* low-level EMFs could adversely affect the human anatomy. It is an odd area to cause such paralysis as we set regulatory guidelines and make health recommendations all the time without understanding such physiological mechanisms. Indeed, the issue hardly even comes up in other areas.

The basic mechanism by which adverse biological effects from EMFs occur is not known at this time, but some likely candidates include:

- EMFs induce internal currents and/or electromagnetic fields, which are then spacially averaged over the whole body.
- Current or electromagnetic fields are induced in critical organs or target sites, which then affect the entire body.
- Certain critical frequencies may exist for specific organs like the brain, or specific cell structures that can influence the entire body.
- There may be a critical time of day when induced currents or fields are capable of producing different effects.
- The earth's static magnetic field may play a role when combined with certain orientations and exposures to external fields.
- There is something crucial we do not know about the human anatomy according to our present model; a radically new model is necessary.

The Human Anatomy

The human body is often likened to a complex machine, but such an analogy does not do us justice, for no machine in existence (now or in the foreseeable future) can function as effectively or efficiently or even approximate the body's complexity. Although the study of illness generally focuses on specific organ systems, in reality almost all activities and disorders affect the entire body. Most of the time, we function through a highly complex internal communications network, with the various organ systems working together in an amazing harmony and synergy on all levels, including biochemical/electrical, intracellular, mechanical, microscopic, and macroscopic.

To illustrate this synergy, consider something as simple as eating breakfast. Long before you awoke and thought about eating, your major organ systems were quietly working at a re- duced basal level. The pulmonary and circulatory systems were providing oxygen and the other nutrients every cell needs. The endocrine and nervous systems were regulating vital autonomic (involuntary) functions like heart rate, blood pressure, and tem- perature control. Metabolism, although reduced, continued dur- ing sleep, involving most body organs. Near dawn, your natural biorhythms speeded up, hormone levels increased, sleep became lighter, and heart and respiratory rates increased.

Hunger may have been one of the first sensations you felt upon waking, because it had been perhaps eight hours since your last meal. The appetite center is located in the hypothalamus re- gion of the brain and transmits its signals through an intricate feedback system involving certain hormones, nerves, and sensory organs. Another immediate sensation you had was probably the discomfort of a full bladder. The kidneys were removing waste and fluid from your blood throughout the night, although in a very concentrated form and at a lowered volume so as not to dis- turb your sleep. (This is why the first urine of the day is darker than at other times.)

Next your major sensory organs came into play, as the eyes, ears, nose, tongue, and mouth began preparing the body for its first meal. The smell of food stimulated the olfactory nerve recep- tors through the mucous membrane linings of the nose, which transmitted the stimulus to the area of the brain where smell is recognized. The brain in turn sent messages through the endo- crine and nervous systems to glands in the mouth and stomach, which then increased the flow of digestive juices. The sounds as- sociated with breakfast were carried to the brain through the outer ear to the eardrum, through the middle and inner ear, to the fine hairlike nerve cells that send the ear's signals to the brain.

Your eyes, too, were involved, sensing light on the lens of the eye and directing it toward the retina, which then transmitted it along the optic nerve to the brain.

With your first bite of food, the nerve receptors on the taste buds sent additional sensory information to the brain. Touch, too, was involved; the various textures of the food helped stimulate the digestive enzymes. Before the food was even swallowed, the enzymes in your saliva began to break down the starches and complex carbohydrates into simple sugars, which are fuel for every cell in the body. All this happened without much conscious effort on your part except for chewing and swallowing.

The involuntary waves of muscle movements called peristalsis move the food through the esophagus and stomach. Most digestion takes place in the stomach, where the food is mixed with gastric juices and more digestive enzymes. The stomach empties into the duodenum, the beginning of the small intestine. But before the stomach empties, a complex interaction takes place between the brain and the intestinal hormones, which are released according to alterations in volume and acid levels. Coordination between the digestive enzymes from the pancreas and bile from the liver and gallbladder is necessary before the proper hormones are stimulated to give the duodenum its alkaline environment. The food's final breakdown occurs in the small intestine, where nutrients are absorbed into the bloodstream.

The digested nutrients flow via the bloodstream from the small intestine into the liver, the body's largest internal organ, which, in concert with the endocrine and other body systems, regulates the amounts of nutrients that enter the bloodstream to nourish other body tissues. The liver controls a number of highly complicated chemical processes, including the production of bile, which is needed for the digestion of fats, and the manufacture of cholesterol, various enzymes, vitamin A, blood-coagulation factors, and complex proteins. The liver also stores various vitamins and minerals, as well as glycogen, which is converted to glucose (fuel) when the body needs it. In addition, the liver detoxifies chemicals, drugs, and alcohol and is the seat of several essential functions of the immune system. Any food that has not been digested moves from the small intestine into the colon, or large intestine, and is eventually eliminated through bowel movements.

While your body was performing this digestive process, other systems were functioning in their own ways. The respiratory system was taking in oxygen and eliminating carbon dioxide with each breath. The circulatory system was carrying the oxygenated blood and other nutrients to all the cells of the body and bringing out the cells' metabolic debris. Almost every cell is in direct contact with a blood capillary. The kidneys were filtering waste and regulating blood pressure and overall fluid balance, while the skeletal and muscular structures were providing strength, overall shape, and movement, as well as protection for the internal organs.

In addition, the marrow of the bones was controlling the manufacture of certain vital blood components, and the endocrine system was generating numerous chemical messengers and hormones, which regulate and influence many body processes, including reproduction and growth. All the while, the immune system has been on the alert for invading microorganisms, and the skin has been carrying on a number of metabolic activities and helping to regulate body temperature. The central nervous system has been providing the information and emotions associated with all of this, as well as coordinating all the activity with the endocrine system.

As complicated as this sounds, it is still an oversimplified picture of what transpires anatomically between breakfast and lunch. Most major organ systems have multiple functions, and no system acts alone, in health or in illness. What affects one system affects all of them. We often forget just how coordinated and precise the human anatomy is. Yet, despite all we know, many things remain mysteries.

Sleep is one of them. We say that the role of sleep is to rest the body, but in theory an organism that is given the proper amount of fuel should not need to rest—energy in equals energy out. And what possible purpose do dreams serve? The healing of wounds is another mystery. What tells the body to decrease its blood flow to an injury? What regenerates the precise kind of tissue that has been damaged? (How does the body know not to regrow liver tissue, say, on a cut fingertip?) What initiates bone growth after a fracture? And then stops the growth once there is enough? Why doesn't the bone just keep growing? And what has gone wrong in the out-of-control cell growth that we know as cancer? Why does it start, and how can we turn it off?

Answers to some of these most fundamental questions may require a different understanding of human anatomy, for no amount of attention to the microworkings of the body is likely ever to explain them. A larger vision is needed.

Circadian Biorhythms

Several theories about the human anatomy and low-level electromagnetic fields have come to the fore in an effort to explain the bioeffects that have been seen. A paradigm shift is slowly occurring because the old construct is just too narrow to resolve the many unanswered questions.

One by-product of EMF research has been a recognition that our present biological models are somehow deficient in their ability to explain certain observed effects. A dual purpose is therefore being served by EMF research: the evaluation of potential adverse affects as well as new ways to understand fundamental processes in living systems that can be put to beneficial use.

The emerging perspective is based on the concept of the human body as a highly coherent biological organism, which is quite different from the microscopic view common to most biologists. The body is, for instance, far more attuned to the various frequencies of the electromagnetic spectrum than anyone had imagined, as our natural circadian rhythms demonstrate.

Circadian ("approximately daily") biorhythm is the term used to describe the body's twenty-four-hour biological clock, which regulates everything from the sleep-wake cycle to our ability to concentrate and perform basic activities, most hormonal activities of the endocrine system, and various functions of the autonomic and sympathetic nervous systems. Even during sleep, these natural biorhythms keep the body functioning smoothly, although in a reduced state of activity. Just before dawn, the natural rhythms increase, appetite is stimulated through the hypothalamus, the heart rate speeds up, blood pressure rises, and hormone levels increase. A new day arrives, even before daylight.

While Eastern practitioners have long associated different times of the day with different organ activities, Western practitioners have just begun to consider circadian rhythms for a variety of purposes, like finding an optimal time of day for various medical treatments. Oncologists, for instance, have found that many patients tolerate toxic reactions to certain chemotherapy treatments better in the morning than in late afternoon.

The biorhythm cycle accepted by some Eastern medical traditions shows these as the times when energy flowing through the various organs is most prominent, although these are not necessarily the times when the activity of the organ itself is elevated. (See chart on the following page.)

Western scientists have also noticed certain twelve-hour rhythms in our biological cycle. We are the sleepiest a few hours before dawn and again at midday, between 1:00 and 3:00 P.M., and are most alert at midmorning and midevening. Such cycles appear true for individual organ systems as well. Liver cells metabolize alcohol the quickest in the early evening and the slowest at 2:00 A.M., hormone levels peak between 7:00 and 9:00 A.M., short-term memory works best between 9:00 and 10:00 A.M., and

1:00–	3:00 A.M.	Liver
3:00–	5:00 A.M.	Lungs
5:00–	7:00 A.M.	Large intestine
7:00–	9:00 A.M.	Stomach
9:00–	11:00 A.M.	Spleen and pancreas
11:00 A.M.–	1:00 P.M.	Heart
1:00–	3:00 P.M.	Small intestine
3:00–	5:00 P.M.	Bladder
5:00–	7:00 P.M.	Kidney
7:00–	9:00 P.M.	Heart constrictor (Pericardium)
9:00–	11:00 P.M.	Triple heater (several organs)
11:00 P.M.–	1:00 A.M.	Gallbladder

all five senses are at their most acute between 7:00 and 9:00 P.M. Biological clocks change according to the seasons, too, with hormone levels rising in the spring and decreasing in the autumn.

Other species have their own circadian rhythms, affected by various natural forces. Nocturnal and diurnal (daytime) creatures have different rhythms; lunar cycles influence mating in some species; and tidal cycles affect appetite and mating behaviors in other species.

Some electromagnetic frequencies are being used medically to help restore biorhythms that have been thrown off balance. Exciting research is being done with the use of light as a therapeutic agent in treatment of depression from seasonal affect disorder (SAD) and of various skin conditions, such as eczema, and as an aid in offsetting jet lag. Light has also been used experimentally to help women normalize overlong menstrual cycles.

Researchers have even begun to look into therapeutic applications for various components of light. Dr. Hugh McGrath, Jr., a rheumatologist at Louisiana State University, has successfully used the UVA-1 frequency of artificial light to mitigate the immune system's attack in lupus patients—especially exciting because of the previous assumption that all sunlight was detrimental to lupus sufferers. Dr. McGrath found that it was only the UVB and UVA-2 components of sunlight that created the typical adverse reactions of rashes, sunburn, and skin cancers. Nine out of ten lupus patients had less joint pain, fatigue, and other symptoms after being exposed to UVA-1 and fluorescent light for ten minutes a day, five days a week, for three weeks. The findings are preliminary, however, since the fluorescent light adds another factor with magnetic properties of its own.

Colors, which are nothing more than light frequencies, have also been found to have varying effects on hospital patients' recovery rates. It's a fascinating subject, to say the least.

Most of the research has focused on light and temperature as the stimulating factors in the daily biocycle. But a handful of pioneers have found that light, while it is a factor, is probably not the determining one. The single most important component of circadian timing is the body's sensitivity to weak electromagnetic fields, and probably to the minute flux in the earth's natural magnetic field within a twenty-four-hour cycle. The implications of this are extremely important because it means we can, and probably are, disrupting our most basic rhythms through externally generated EMF exposure.

In studies spanning more than twenty-five years and hundreds of subjects, Dr. Rutger Wever, of the Max Planck Institute for Psychiatry in Germany, found that in a constant shielded environment, without clues as to the time of day, people adopted a twenty-five hour or longer day, no matter what the light conditions were in their chambers. In various animal experiments, he had found light to be the most effective stimulus of circadian rhythms, but humans seemed resistant to light waves alone as a circadian stimulus. What he found was that the standard triggering stimulus in humans was a weak ELF field of 2.5 volts per meter at the 10-hertz frequency. That is within the range of the micropulsations of the earth's natural background and is also the frequency of most EEG brain-wave activity in animals and humans.

Other Frequencies and How They Affect Us

The human central nervous system functions through minute pulses of electrical energy, typically in a few millionths of a volt, and it doesn't take a lot of outside electrical interference to disrupt such tiny signals. The heart depolarizes (shifts from positive to negative current) and produces its beat in only about a quarter of a second. The sensation of pain can be felt at currents as low as 10 milliamperes on the skin; ranges of around 200 to 300 milliamperes can be lethal.

In experiments with humans reported many years ago, frequencies at 6.6 hertz were found to cause depression; 11 hertz caused agitation and rioting; 8 hertz created feelings of elation; and frequencies below 6.26 hertz caused confusion and anxiety. Also, green, blue, and red color perceptions were intensified when electrodes were placed on the temples and people were exposed to low voltages at 42.5 and 77 hertz. Some studies found that certain fields increased general visual sensitivity but reduced color perception.

Brain waves have long been known to have specific frequencies: delta waves, which are associated with deep sleep, are between 1 and 3 hertz; theta waves, which reflect mood, are between 4 and 7 hertz; alpha waves, which signify relaxation, are between 8 and 12 hertz; and beta waves, which are linked with conscious thought functions, are between 13 and 22 hertz.

Pulsed lights are also known to alter human behavior; 6 to 7 hertz can induce anger; 10 hertz is usually soothing, except in persons prone to epileptic seizures (which may be triggered when the frequency synchronizes with the brain's alpha waves). Some susceptible individuals have even become involuntarily violent in movie theaters from the flickering of the old twenty-four-frames-per-second, 16-millimeter films.

The National Institute for Neurological Diseases found that a frequency of 388 megahertz was lethal to monkeys, and further experimentation determined that a sweep of 380 to 500 mega-hertz caused pulsing in the brain, ringing in the ears, and reactions of rage in human test subjects. Most interesting was the fact that each individual appeared to have his or her own resonant frequency, which researchers speculate is a by-product of a person's body height serving as a half-wave antenna. (According to tests on humans, we absorb energy most efficiently at around 82 to 87 megahertz, which is the frequency range used by many FM devices, but we resonate—respond as an antenna—right in the middle of the UHF television band.) Humans have been found to react to radio transmissions at 129 megahertz, and some bands in the VHF ranges are known to induce hallucinations. Also, UHF frequencies have been found to change the optical properties of glycogen, which provides energy to the muscles.

Decades of animal studies have shown a variety of effects. UHF frequencies altered brain-wave patterns in rabbits, reduced conditioned reflexes in rats, and changed the heartbeat of chick embryos. In short exposures, UHF frequencies increased the re-growth of severed tissue; long-term exposures repressed it. Some weak fields proved analgesic; stronger fields induced pain. Expo-sures at 29 megahertz have been used to kill insects in bread. The swaying dance of the cobra corresponds to a 3-hertz field and is thought to hypnotize its prey. And most animals panic at vibra-tions in the 7-to-15-hertz range common to earthquakes.

One noted researcher, José Delgado, a neuroscientist at Centro Ramon y Cajal hospital in Madrid, Spain, has reported amazing results with various forms of EMFs in the myriad species he has studied. They include flashy experiments in the 1960s when he stopped a charging bull dead in its tracks with a radio-frequency signal sent to an electrode buried in the bull's brain;

experiments with chick embryos exposed to very weak fields, which produced gross deformities; and similar studies that produced lethal genetic mutations in fruit flies. In all this work, Dr. Delgado has found many adverse effects with exposures several hundred times below the voltage required for an electrode to stimulate the firing of a nerve cell.

Dr. Delgado and other researchers have consistently noted what is called the entrainment phenomenon. When he exposed crab neurons that were already firing at a specific rate to particular fields, the neuron changed its firing rate to synchronize with the applied pulse. Dr. W. Ross Adey, a neuroscientist at the Veterans Administration Hospital in Loma Linda, California, has likewise found that the brain waves of monkeys lock into phase with externally applied fields tuned to the same frequency band as the brain waves. This entrainment phenomenon can be thought of as the electromagnetic equivalent of the way a mother's and child's breathing rates synchronize when they sleep together.

Some say that none of the behavioral research on primates and humans has been missed by the military, that they have been developing nonlethal EMF weapons for years that make use of various frequencies to control human behavior. There is reason to believe that such weapons were used in the Gulf War. The military denies this but acknowledges that top-secret research on the subject continues.

Some experts even speculate that given the rage reactions noted in some studies, there may be a link with the increasing violence in our society, and an ever-increasing ambient EMF background across a wide spectrum of frequencies.

Central Control:
The Pineal Gland and Melatonin

How weak ELF fields can affect the human anatomy, keep in mind, isn't understood. But some think the connection will be found through the intricate ophthalmic nerves, which are light sensitive, and the ethmoid/sphenoid sinus region, with its rich concentration of magnetite around the eyes and sinuses, as well as the magnetite in brain tissue. Others think the connection will be found in the electrical properties at the surface of cells. Still others believe that the human anatomy uses electromagnetic information in a complex way that indicates the existence of an alternate central nervous system.

In experiments on visual acuity, by Cremer-Bartel in 1983, weak ELF fields were found to affect a specific enzyme needed for the synthesis of melatonin, the "master hormone" that seems to affect everything from the circadian rhythms to the production of seratonin, an important neurotransmitter.

Melatonin is produced in the pineal gland, a small pinecone-shaped gland located deep within the center of the brain. Not so long ago, physiologists considered the pineal gland, which is about the size of a pea, to be as useless and dispensable as the appendix. The pineal gland has since turned out to be perhaps the most important gland in the body. Dr. Robert Becker speculates that it might prove to be the human equivalent of a "magnetic organ." Researchers are increasingly surprised at the extent of the physiological processes that are either controlled or influenced by the melatonin the gland produces. Among them:

• Melatonin production in humans is elevated at night. Production is stimulated by ambient darkness and suppressed by bright light. When levels are high, people feel drowsy. The various experimental light therapies are thought to succeed by altering melatonin production. The direct ingestion of melatonin is being tested as a possible treatment for insomnia, jet lag, seasonal depression, and some forms of cancer, with tantalizingly positive results.

• Melatonin secretion decreases over a lifetime, peaking in childhood and gradually lessening after puberty. People over sixty secrete far less than do twenty-year-olds—perhaps a clue to the cause of age-related insomnia and some degenerative diseases.

• Melatonin is one of the most efficient destroyers of free radicals in the body. Free radicals are naturally occurring unstable oxygen molecules that can play a key role in the development of atherosclerosis and heart disease, as well as various forms of cancer, because of the damage they do to DNA, cells, and tissue. Alzheimer's patients, those with Down's syndrome, and sufferers from clinical depression have all been found to have low levels of melatonin.

• Melatonin has natural anticancer properties, including an ability to increase the cytotoxicity of the immune system's killer lymphocytes. If melatonin is lowered in the body, various cancers may proliferate; decreased melatonin levels have been implicated so far in breast cancer, prostate cancer, melanoma, and ovarian malignancies.

• Melatonin therapies in some laboratory studies have shown positive results against breast cancer in reducing the number of tumor-producing estrogen receptor sites within a malignancy, the size of tumors, and reoccurrence. It is being experimentally combined with Tamoxifen (an estrogen inhibitor) in a few human clinical trials. Some speculate that it is not melatonin's ability to reduce estrogen receptor sites that accounts for the promising results, but, rather, its ability to suppress surges of other hormones that can promote breast-cell growth.

• Melatonin's relation to light is evident in the fact that blind women typically have higher levels of it than do sighted women. Also, the incidence of breast cancer is far less in blind women.

• Melatonin production is influenced by electromagnetic frequencies other than those of light. Studies have found that melatonin suppression through the pineal gland occurs at frequencies not far above those of the earth's natural magnetic background and within the scope of the common household ranges of 50 to 60 hertz. If melatonin's production is suppressed at night, due to particular EMF exposures from electric blankets, waterbed heaters, or ambient background exposures, health repercussions may be especially severe.

The incidence of breast cancer is steadily increasing in most industrialized countries. Several researchers today speculate that the increasing use of electricity and the increase in breast-cancer incidence are directly related. If weak ELF fields suppress the pineal gland's production of melatonin, as well as the immune system itself, and if there is a high correlation between low melatonin levels and breast cancer, the relationship seems obvious.

In five studies, elevated electromagnetic fields have been implicated in an increased incidence of male breast cancer, a very rare form of the disease. Men who worked as telephone linemen, in switching stations, as electricians, or as electrical workers in the utilities industry were found by some studies to have a sixfold increase in breast cancer. Recent findings have verified an increase in deaths from breast cancer among women in telephone and electrical occupations. Epidemiological surveys have also found elevated rates of breast cancer in women who live near high-tension lines. All this should raise another of those red flags on the playing field. The subject requires immediate attention since so many women are occupationally exposed to EMFs through office work alone. Simple shielding devices and inexpensive mitigation techniques may in time save thousands of lives.

Melatonin suppression may supply researchers with a key piece of the puzzle as to how EMFs could be associated with so many different forms of seemingly unrelated cancers, breast cancer among them.

Other Hormones and Neurotransmitters

Hormones are chemical coordinators or initiators that produce an effect in some area of the body, often remote from the gland in which the hormones originate. Although they do not constitute a definite organ system (like the cardiovascular system), the glands that secrete hormones are grouped together in what's called the endocrine system. Hormones use the blood-stream, rather than their own specialized ducts, to communicate their messages.

Neurotransmitters are a special class of hormones that carry nerve impulses from the brain across cell membranes throughout the body. Seratonin, dopamine, epinephrine (adrenaline), and acetylcholine, among others, are neurotransmitters. They interact together, sometimes as if in units. One researcher has likened it to the cells "whispering together" to transmit the body's most important messages through the complex workings of the endocrine and central nervous systems.

The importance of neurotransmitters is a relatively recent focus of research. Low seratonin levels, for instance, have now been implicated in violence-prone and suicidal behavior. Known suppressants of seratonin include liquor, some drugs, stress, and lowered levels of melatonin, as well as electromagnetic fields. Some European researchers have puzzled over clusters of depression and suicide among people who live or work near high-tension lines or in environments with very strong fields, like those near radio, TV, or radar transmission facilities. They wonder if low seratonin levels may eventually be found to be the cause.

EMFs alter the production of neurotransmitters in a way that has serious implications for cancer and other diseases. Among other functions, neurotransmitters bind and interact with receptor sites on the surface of the cell. These receptor sites tell the cell what to do; they are the docking portals through which information travels from the outside to the inside of the cell. The number of receptors can vary greatly, and researchers often gauge how much activity is going on within a cell based on the number of receptors it has. The number of estrogen-receptor sites on breast-cancer cells is used in determining whether estrogen blockers might be recommended, for instance. Certain chemicals are known to increase the number of receptor sites, and studies have shown that EMFs also greatly increase them. Some researchers think that EMFs may open up the basic cell structure to a host of changes that cannot be regulated. It is one of several interesting areas of current EMF research.

Stress Response

Another important aspect of EMFs in relation to the endocrine system involves a more general physiological response. There is ample evidence in test animals that EMFs increase the levels of adrenaline, the fight-or-flight hormone released from the adrenal glands, which are located on top of the kidneys. Stress is primarily mediated through the adrenal, pituitary, and thyroid glands, and plasma cortisol is a substance produced by the adrenal glands during stress conditions. Test animals continuously exposed to high-frequency microwaves of 2.45 gigahertz at about 0.5 milliwatts per square centimeter (0.5mW/cm2), approximately twenty times lower than what is considered a safe thermal exposure, were found to have a fourfold increase in cancers of the above-mentioned glands and increased levels of plasma cortisol. Benign tumors of the adrenal glands called pheochromocytomas, which can cause the chronic increase of blood pressure to a dangerous level, also showed a significant increase in test animals.

Similar studies investigating EMFs and stress present several interesting paradoxes. Test animals appear not to know they are stressed, yet blood tests show high levels of cortisone, a substance released in the body under conditions of long-term disease, as opposed to adrenaline, which is released in a fight-or-flight response. Monkeys exposed to a 200-gauss magnetic field for four hours a day showed a generalized stress response for six days, which then declined, suggesting that the animals had adapted to the exposure. Researchers who stop experiments at that point can reasonably conclude that there has been no long-term damage. However, in subsequent experiments, it has been found that when the exposure continues, hormone and immune levels will fall far below normal and remain there. The immune system becomes exhausted and unable to rebound, opening the body to infectious diseases and an inability to fight malignancies.

Russian studies in the 1970s found that rats exposed very briefly to even small amounts of microwaves released stress hormones; the same results have been found at the 50-hertz frequency. Other studies found an exhaustion of the adrenal cortex with exposures to 130-gauss magnetic fields at the 50-hertz frequency. One thorough Russian biophysicist, N. A. Udintsev, found not only the slow stress response but also an adrenaline release in test animals exposed for just one day. Hormone levels did not return to normal for nearly two weeks. He also found a rise in blood-sugar levels and an insulin insufficiency at the same frequencies. Russian research has repeatedly reported high blood pressure and cardiac irregularities in humans exposed to microwave frequencies.

In 1976, J. J. Noval, at the Naval Air Development Center in Johnsville, Pennsylvania, solved the puzzle of how high-stress chemicals could be present in animals who did not appear to be feeling stressed. Noval found the same slow stress response that others had observed at very weak electric fields of around 5 thousandths of a volt per centimeter, but he determined that when this vibrated in the ELF ranges, the level of the neurotransmitter acetylcholine increased in the brain stem. This occurred in a way that sent a subliminal stress signal throughout the body without the animal's being aware of it.

The implications of EMFs in relation to subliminal stress are important for several reasons. Stress is often thought of as a purely emotional state, but it is also a chemical one and creates a whole cascade of chemical responses in the body. Prolonged chronic stress is detrimental to every anatomical system, including the reproductive one. Subliminal stress may affect fertility and elevate blood pressure, which can lead to heart disease and

strokes, as well as suppress immune function. People may be unaware of these chemical alterations. Even short EMF exposures,
like the use of a cordless phone on and off throughout the day,
could cause spikes in such hormone levels. More tests need to be
done on a 24-hour basis that truly represent the way we interact
with the various fields we are constantly exposed to.

In addition, female test animals showed hormonal stress responses (such as fur discoloration) to magnetic fields, which male
animals did not show. It is known that women react differently to
stress than men do. What are the implications of this for infertility and reproductive problems in female office workers, for example?

A New Model: The Body's Own Direct Current

Dr. Robert Becker, building on a lifetime of professional research, has developed one of the more interesting broad-view
hypotheses to encompass the many enigmas inherent in the study
of electromagnetic fields and biological responses. His theories
bring energy back into living biological systems, not in the former
guise of vitalism and its accompanying superstitions, but rather
as the organizer and controlling agent for the total organism and
its most important functions.

Dr. Becker believes that science now supports an emerging
concept of a dual nervous system, a kind of direct-current, analog
system that is different in overall purpose, function, and ability
from the nerve-impulse system that we had presumed to be our
only one.

What led Dr. Becker to this conclusion were several ground-
breaking findings over many decades, through his own research
and that of a handful of such like-minded colleagues as Drs. Andrew Marino, Andrew Bassett, Maria Reichmanis, Charles Bachman, and Joseph Spadaro. Dr. Becker is perhaps best known for
his work with salamanders and limb regeneration, which has

been expanded and applied to human bone growth through electrical stimulation. (Such electrical applications are now being used to heal bone fractures that will not fuse through the body's normal means.)

In his two books, *The Body Electric*, written with Gary Selden, and *Cross Currents: The Perils of Electropollution, the Promise of Electromedicine*, Dr. Becker sets out these hypotheses:

• The body has a distinct direct-current system, not unlike the ancient meridian acupuncture system of Eastern medicine. In fact, Dr. Becker and Dr. Reichmanis investigated the electrical properties of the traditional acupuncture points for the National Institutes of Health. Not only did they locate at least 25 percent of the acupuncture points on the forearm, but they found that each point had specific reproducible electrical parameters in all test subjects. They also explored the meridians that are said to connect these points and determined that they had electrical characteristics not unlike transmission lines, whereas nonmeridian skin did not. They hypothesized that the acupuncture points serve as booster amplifiers along the meridians to keep electrical current moving. Since acupuncture has been successfully used for centuries to block pain, inserting metal needles at specific points would be expected to create enough interference with the electrical properties to halt the flow of information in an internal DC system. The U.S. military and the Soviets have both investigated the use of acupuncture as an effective battlefield anesthesia. Other researchers have verified the electrical properties of acupuncture points and have measured current in the meridians.

• The internal DC system is responsible for the way the body initiates wound healing. While we have developed a sophisticated understanding of the process of tissue repair, no one has been able to locate the switch that turns the process on and off. Dr. Becker, in his early work with salamanders, discovered an interesting phenomenon. Salamanders have the ability to completely regenerate severed limbs, eyes, ears, the digestive tract, half of the heart, and a third of the brain, making the creature—in theory, at least—almost immortal. What forms in this regeneration is an exact replica of what was there: bones, nerve fiber, blood vessels, muscle, and so on. Frogs, on the other hand, while similar to salamanders in many respects, cannot regenerate anything beyond the healing of bones, like humans. Dr. Becker wondered if there was some inherent difference at the wound site between the two amphibians that could initiate regeneration in one but only scarification-type wound healing in the other. What he discovered was that the two amphibians had different electrical properties at the injury site. The amputation site of the salamander gave off an electrical DC current that was negative in polarity, while the amputation site of the frog was positive in polarity. Dr. Becker's work over several decades reintroduced electromagnetic energy back into biology as a controlling factor in healing processes. What he measured was the current of injury Galvani had first observed in the late 1700s.

• Embryonic cells do not just grow forward in progression; they also grow backward—a way of "rewinding the growth tape." In his regeneration work, Dr. Becker noticed that the first thing that happened after salamander limb amputation was a rapid growth of outer cells of the skin across the surface. Next, the nerves at the cut ends of the stump connected with each skin cell, which is called the neuroepidermal junction. Soon, a mass of primitive embryonic cells called the blastema formed, out of which the new limb would grow. Originally thought to be new cells, it turned out that the primitive cells forming at the injury site were in fact old mature cells from bone, muscle, and so on that had returned to their embryonic state. The process is called dedifferentiation, and it was considered medical heresy at the time. (Dedifferentiation is now known to be true, and its importance will be discussed in the next chapter as there is evidence that cancer cells are in a state of dedifferentiation.) What Dr. Becker and his collaborator, Dr. David Murray, chairman of the Department of Orthopedic Surgery at the State University of New York Medical School, dis-

covered was that red cells in frogs could be made to dedifferentiate by means of electricity in vanishingly small amounts
measuring in the billionths of amperes. Electricity was fundamental to primitive cellular processes.

• All cells are not electrically alike. There must be the right
amount of electricity of the correct polarity, as well as cells sensitive to DC electrical currents, for effects to occur. But a larger,
more coherent system must also exist. Input DC electrical signals
carry injury information along the acupuncture meridians to various parts of the brain, which in turn switches on output DC signals to activate the chemical processes involved with injury
repair. Dr. Becker envisions a complete closed-loop DC system
within the body, which is connected to the central nervous system
as we know it.

• The internal DC system forms a kind of template around and
throughout the body. We exist in our own evolving morphogenic
field, which contains all the organizational information needed to
tell what to re-form, and heal, and where. The primary frequency
of EEG brain waves in all animals is within the 10-hertz range,
which is an extremely low frequency but remains an essential one
for a great deal of biological activity. Dr. Becker, psychiatrist How-
ard Friedman, and physicist Charles Bachman measured the DC
electrical activity in the human brain and found that back-to-
front DC current varied with changes in consciousness. It
strengthened during mental activity and physical exercise, declined during rest, and reversed its potentials altogether in regular sleep and under anesthesia.

Polarity reversals during sleep may provide a clue to sleep's
basic function in our lives. It may be the time when the body is
rejuvenating itself, cleaning up its cellular debris, checking on
and regulating its most basic functions, and making cell repairs.
Anything that disrupts sleep, such as an interference with melatonin levels or circadian rhythms, may have implications beyond
mere tiredness the next day.

There is far more to Dr. Becker's work and writings than can be explored here, but his books are available to anyone interested in knowing more.

Like most pioneers, Dr. Becker has paid a high price for a vision that challenges popular orthodoxies. He has had untold grants awarded and then pulled, and has suffered the scorn of ideologues within the scientific community. But his theories will probably prevail in the end, for the simple reason that they are the only ones expansive enough to offer plausible explanations beyond the point at which the chemical-mechanistic model utterly fails. The reintroduction of electromagnetism's influence on our fundamental biological processes may be one of the most important medical breakthroughs of the century, and is a very exciting way to enter the new one.

Chapter 8

THE SECRET ELECTRICAL LIFE OF CELLS

THE HIGHER up the evolutionary ladder and the more developed the nervous system of any living organism, the more complex are its electrochemical interactions and properties. What occurs between the light-sensitive pineal gland's release of melatonin, and other hormones that communicate complex signals through the bloodstream, helps to determine everything from basic growth to heart rate, blood pressure, fluid content, and fundamental cell changes. Clearly this is far more involved than what transpires in a one-celled organism like a paramecium, even though its basic functions are complex in their own way too.

The smallest unit of all biological activity is the cell. Cells are packed with highly charged atoms and molecules that can change their orientation and movement when exposed to certain electromagnetic fields. What happens on the cellular level influences the entire organism, and vice versa.

Cells and How They Grow

One of the most important attributes of cells is their ability to grow, differentiate, and reproduce, and electromagnetism is involved in much of this cellular activity. Every cell membrane has an electrical charge across it that moves from the inside to the outside of the cell, and several key phases of cell division are electrically stimulated.

Cells come in many sizes and shapes. Some are so small they cannot be seen with a normal microscope; others measure inches across. A nerve cell, for instance, can be three to four feet long and as thin as a thread. And there are exceptions to nearly everything that can be said about cells. Not all of them have a nucleus; some muscle cells have several, but red blood cells have none. Here, however, we will generally be speaking of the basic cell, which consists of an outer membrane, an inner membrane, a nucleus, and the content of the nucleus.

In its normal environment, the living cell is in dynamic equilibrium with everything around it. It maintains its characteristic form and functions even while in a state of constant change, taking in various nutrients from outside and discharging others. The body's enzyme system fuels this continuous metabolic process, but it gets its instructions from the nucleus of the cell itself. Anything that confuses this delicate, complex information network threatens the cell's precision at its most elemental level.

The cell's nucleus contains most of the body's hereditary information in the chromosomes and the genes arranged in strands along the chromosomes. Genes are composed of double strands of DNA (deoxyribonucleic acid) coiled into a twisted helix. DNA controls most cellular activities through the synthesis of protein. It uses single-strand RNA molecules, which the DNA synthesizes, to transfer information across the cell's cytoplasm.

Depending on its function, there are several forms of RNA. The formation of messenger RNA from DNA is called transcription; the synthesis of protein by messenger RNA is called translation; the duplication of DNA is called replication; and at least one other form of RNA, called ribosomal RNA, is probably involved with protein synthesis as well. A host of amino acids and enzymes also play a role in the complex life of the cell.

Normal Cell Division and EMFs

Cells divide and reproduce in a continuous process called mitosis, which begins with a single cell that grows into two independent ones. It occurs in the nucleus and is basically the duplication and equal distribution of the chromosomes into daughter cells. Not all cells undergo mitosis. Mammalian red blood cells, which have no nucleus, cannot divide, nor do such highly specialized cells as neurons. But some cells undergo mitosis often. The embryo is one, and that is why exposure to EMFs is of particular concern during pregnancy and a child's early growth years.

The illustration, "Phases of Mitosis," shows what happens. Mitosis, which takes only minutes to complete, has four phases: prophase, metaphase, anaphase, and telophase. The period between divisions is called the interphase, or resting phase, which is the longest period (about twenty-four hours). But resting is somewhat of a misnomer, since a lot is transpiring then, including DNA duplication.

In the prophase, chromosomes begin to appear out of the DNA mass and are found to have doubled in number. The membrane around the nucleus disappears. Spindle fibers also appear and extend toward the midpoint, or equatorial plate, of the cell. Their orientation will determine the direction toward which the chromosomes will move, as well as where the cell will pinch together to form two cells.

In the metaphase, the chromosomes line up along the equatorial plate at mid center. During the anaphase, the chromosomes separate in the directions established by the spindle fibers, half going toward one pole and half toward the other.

In the final stage, telophase, the cell pinches in all around until two daughter cells have formed. At the end of the telophase, a nuclear membrane forms on each new cell, and it usually returns to the interphase state.

We can view this happening under a microscope, but several questions remain unanswered. What signals mitosis to begin?

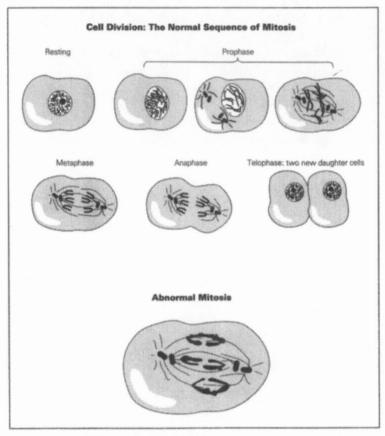

Cell Division: The Normal Sequence of Mitosis

Resting

Prophase

Metaphase

Anaphase

Telophase: two new daughter cells

Abnormal Mitosis

Resting phase: DNA is being duplicated.

Prophase: Chromosomes and centrosomes appear, match up and begin to move to the cell's center.

Metaphase: Chromosomes line up in pairs along the cell's center line.

Anaphase: The spindle fibers of the centrosomes recede and pull the two chromosome pairs apart.

Telophase: Two new daughter cells have been formed, each containing exactly one of the original chromosome pairs.

Abnormal Mitosis: During the anaphase, abnormal bridging between chromosome pairs can result in unequal amounts of chromosomes going to each new daughter cell. Chromosomes have difficulty pulling apart. Faulty genetic information in a cell can create oncogenes, as well as a host of other abnormalities.

What fuels the process? What actually pulls the chromosomes away from each other? What causes the cell to pinch at the end? What significance does this have for cancer, which is a sort of uncontrolled mitosis? And what about EMFs?

There is clear evidence that all cells involved with active division are affected by the extremely low frequency (ELF) ranges, as well as by the microwave band. Some think a resonance phenomenon (about to be discussed) may be found to link mitosis with the earth's natural magnetic fields. The presence of abnormal artificial EMFs has already been implicated in disturbances to normal mitosis and in genetic abnormalities.

Dr. Abraham Liboff, of Oakland University, found that mitosis was speeded up when cells were exposed to ELF magnetic fields. Dr. Martin Poenie, of the University of California at Berkeley, found major changes in calcium ions during the anaphase stage, and several researchers have linked calcium ion changes with EMFs, as well as with the cyclotron-resonance phenomenon, raising the possibility that low-level exposures to EMFs may be altering the genetic process and causing abnormal cell growth.

Studies have shown that the exposure of cells to 27-megahertz frequencies was followed by an abnormal bridging between the chromosome pairs during the anaphase (see the illustration). Matching chromosomes had difficulty in separating, which meant that the two new cells did not get equal amounts of genetic information. Oncogenes, which cause cancer, can theoretically be formed under such circumstances. Other studies appear to confirm this.

Calcium Ions

Cells are tiny, dynamic units which maintain themselves by extracting information, materials, and energy from their surrounding environments. The information "currency" they use between their vigorous internal workings and their "outside" environment is calcium.

Calcium ions, stored in the cell membrane, are charged particles that play an important role in numerous cellular processes, including the firing of nerve cells. They are released when hormones transmit instructions to the receptors on the cell's surface. The calcium ions are the primary messengers of cellular activity, relaying vital signals from hormones, antibodies, and neurotransmitters back and forth across the membrane. They are like a chain of dominoes, capable of transmitting various perturbations into the interior of the cell. In addition, calcium mediates various signals between cells that prevent unregulated cell growth.

Calcium therefore exerts a crucial control over cell function, growth, and differentiation, including such pathological processes as tumor formation. Ion channels exist between some cells—possibly this is the way the "whispering" cells communicate, although there is a coherency to their communication that defies simple channels. It has been found that fields too weak even to stimulate nerve impulses can nevertheless change the way calcium ions bind to the cell surface, thereby unleashing a host of chemical reactions deep within the cell.

Several studies have found that certain critical frequency windows release calcium from cells. The phenomenon is called the calcium efflux response. In 1975, Suzanne Bawin and W. Ross Adey, at the Pettis Memorial Veterans Medical Center in Loma Linda, California, found that the brain tissue of newly hatched chicks released calcium ions when exposed to a radio-frequency field modulated at 16 hertz, but not at 1 or 30 hertz. This indicated the possible existence of critical frequency windows for all species. The effect has been repeatedly observed since in human brain cells in laboratory cultures and in numerous animal studies.

Dr. Carl Blackman, a research biophysicist at the Environmental Protection Agency, found important windows for calcium "dumping" at 15, 45, 75, 105, and 135 hertz. The effect turned off at 30 hertz and at other frequencies under some circumstances. Dr. Blackman and others expanded the research to factor in the earth's geomagnetic field in the range of 38 microtesla, experimenting with it in varying ratios. They found that at different strengths, the earth's magnetic fields played a role when combined with the externally generated frequencies mentioned above. Many deduce from this body of work that living organisms are greatly influenced by their interactions with the natural electromagnetic background, humans included. It also adds an extra layer of complexity to any subsequent EMF investigations, since a net static field was required in combination with the geomagnetic field to produce the window effects that were observed. A resonance phenomenon is probably responsible.

The earth's background magnetic field is generally between 0.2 and 0.6 gauss, but the strength varies in different locations and at different times of day. Such fluctuations influence our circadian rhythms, and therefore our metabolic processes, in ways we have only begun to discover. But this work may point in particular to a theory of the suppression of melatonin *at night* (when levels are supposed to be high) in some as-yet unknown relationship with the naturally occurring change in the earth's geomagnetic field at that time, creating an adverse augmentation at the cellular level. In other words, the time of EMF exposures, such as the use of electric blankets or sleeping with one's head too close to an analog clock during the night, may be an important factor along with considerations for frequencies, modulation, and duration of exposures.

A Lethal Triad

Changes in cellular calcium flow stimulate an important group of enzymes called protein kinases, which play a significant

role in regulating several cellular functions, including cell proliferation itself. Decreased protein kinase activity in human
lymphocyte cells in the immune system have been reported in
several EMF studies. Researcher Daniel Lyle found the immune
system's killer-cell activity was inhibited by up to 10 percent upon
exposure to some fields. Other studies found T-lymphocyte suppression up to 20 percent. This means that the immune system
itself is being compromised at a time when chemical messengers
may be giving erroneous instructions to internal cellular regulation systems.

Another important enzyme involved with cell growth is ODC
(ornithine decarboxylase), which is always present during cell
growth. But increased levels of ODC are considered a marker for
the kind of heightened cell activity common in cancers. Research
by Craig Byrus has found that EMFs at 60 hertz turned on ODC
production when combined with certain chemicals known to be
carcinogenic, but not without them. This indicates that EMFs
are co-promoters with other carcinogens.

Paralleling this work were studies by researcher Jerry Phillips
done on human colon-cancer cells in laboratory cultures. The
cells were found to become more tumorigenic when exposed to
magnetic fields.

Taken together, what this body of work indicates is that at the
cellular level, EMFs induce ODC activity, which in turn activates
cell growth at a time when the immune system is slightly suppressed, and tumors themselves have been stimulated into rapid
growth. This is a lethal triad of events. Not all of these studies
have been replicated, however, and a coordinated series of like
studies at the same frequencies is called for.

Another disturbing body of research related to cell growth is
that done over a fifteen-year period by German scientists Grun-
dler and Keilmann. A type of yeast cell commonly used in biomedical studies to indicate cellular growth and genetic mechanisms
was found to have a considerably enhanced or reduced growth

rate, depending on exposure to different frequencies within the microwave bands. The cells also showed marked deformities.

Cells and Resonance

We have made passing reference to another important theory about how it is possible for vanishingly small amounts of externally applied energy, when combined with the earth's own fields, to produce major biological effects beginning at the cellular level. It's a bit complicated, and you might have to refer back to the physics chapter in order to understand the relationships about to be described. The theory explores a concept called resonance phenomenon, cyclotron resonance in particular.

There are several forms of resonance, the best known today being magnetic resonance imaging, or MRI, which is used diagnostically like X-ray. Cyclotron resonance is a complex phenomenon based on the relationship between a constant magnetic field and an oscillating (time-varying) electric or electromagnetic field that can affect the motion of charged particles such as ions, some molecules, atomic nuclei, or electrons in living matter.

The body is filled with charged particles, and many of our most essential biological processes involve any number of them, including sodium, calcium, lithium, and potassium ions, all of which function by acting on or passing through the cell membrane.

Cyclotron resonance is what occurs when an ion is exposed to a steady magnetic field that causes it to move in a circular orbit at a right angle to the field. The speed of the orbit is determined by the charge and mass of the ion and the strength of the magnetic field. If an electric field is added that oscillates at exactly the same frequency and that is also at a right angle to the magnetic field, energy will be transferred from the electric field to the ion, causing it to move faster. The same effect can be created by applying an additional magnetic field parallel to the constant magnetic field.

The reason this is important is twofold: one, it provides a plausible mechanism for how we interact with both normal and abnormal fields; and two, it explains how vanishingly small amounts of externally applied energy, when concentrated on specific particles like sodium, calcium, potassium, or lithium ions, can cause major biological activity. It also points to the body's ability to extract (demodulate) certain pieces of electromagnetic information from the natural environment and make use of them, through the pineal gland and other organ systems.

The importance of resonance theory should not be underestimated. Keep in mind that resonance is a *relationship* between a steady-state magnetic field and an oscillating electric or magnetic field. The resonance relationship applies to all frequencies within the spectrum and as such is not based on power density alone. What's required to create resonance is the right combination of things. For instance, Drs. John Thomas, John Schrot, and Abraham Liboff, working at the U.S. Naval Medical Research Center in Bethesda, Maryland, produced resonance in hypothetical lithium ions in the brain tissue of rats when the rats were exposed to a controlled magnetic field of 0.2 gauss (equivalent to the lower range of the earth's magnetic field) and 60 hertz. In addition, the test animals' behavior changed measurably.

There are other forms of complex resonance, like nuclear resonance and electron paramagnetic resonance, which are also important possibilities but are too complicated to do justice to here and are less well studied besides.

What cyclotron resonance signifies for the human anatomy is this: The earth provides a natural steady-state magnetic field of between 0.2 and 0.6 gauss. The frequencies for the oscillating fields required to produce cyclotron resonance in some important ions in the human body fall within the ELF regions of between 1 and 100 hertz. This is the frequency range of all our electric-power generation in the 50-to-60 hertz range. Complex resonance may account for many of the adverse reactions reported around high-tension lines, for instance. But the ELF frequencies are not the only ones that are cause for concern. The human anatomy can apparently demodulate, or extract, many frequencies, including the microwave bands that constitute many of our communications frequencies. Since resonance is frequency dependent, *not* power dependent, safety standards that are primarily based on power-density factors and tissue-heating models, like the ANSI standards, are meaningless.

That humans interact so fundamentally with the earth's magnetic fields many now think may be the most important biological discovery of this century. It could be a constant hidden variable in all scientific research, too.

Differentiation

Multicellular organisms, which include just about everything above bacteria on the evolutionary ladder, require a process called differentiation in order to keep the whole biological orchestra playing in tune. The more developed the species, the more complicated this process becomes.

Differentiation is an intracellular process that suppresses some DNA information so that whatever information is needed at a given time will predominate. The DNA of each individual cell contains all the genetic information the body will ever need. But in order for the body to manufacture eye tissue, say, or liver tissue, muscle or bone cells in the places where they belong, something must direct each cell on what to do; otherwise, growth would be utter chaos, not to mention antithetical to life itself. That is what differentiation does.

An embryo—be it a mouse, a bird, a fish, or a human—begins as a single-celled fertilized egg and grows into a tremendously complex living being, comprised of literally billions of various cells. We have described the *mechanism* of differentiation as the repression of all the DNA in a cell except that required for a specific cell type. But cells are not just moronic lumps of dividing tissue. There is an organizing principle behind what eventually becomes the living entity that a mere description of the process misses completely.

What tells the cells to form muscle or bone in a particular place or to develop the specific organs? And then what tells them how to coordinate all of this with the brain and the central nervous system? Dr. Becker (see previous chapter) has theorized that we have a closed-loop control system that not only directs differentiation from the fertilized egg all the way to mature growth, but also directs injury repair (with specific on-and-off signals) and general cell replacement. He believes that such a system does not require either intelligence or conscious thought as we know it and is "indistinguishable from life itself." Such a negative-feedback control-loop system appears to be present in all living things and is probably electrical in nature. It is also a system that might be enlisted to help when health is deteriorating or cancer cells are growing out of control.

Cancer: Abnormal Cell Division and EMFs

As mentioned before, cancer is a state of uncontrolled mitosis in which cells randomly divide and grow after escaping the body's normal control mechanisms. One doctor described cancer cells as "wilding teenagers." That metaphor is a particularly appropriate one given what molecular biologists have come to understand about cancerous cells. While we often think of cancer cells as foreign invaders separate from "us," in fact, they are our normal cells that have become stuck in a state of partial dedifferentiation.

Scientists used to regard differentiation as a one-way street, with a cell capable of going only from an embryonic state to a mature one. It is now known that this process can and does reverse itself. For example, in wound healing the cells at the wound site must return to the embryonic state in order to heal.

With cancer, the defect is in the cell itself, not in the body's overall systems. Agents that initiate cancer (viruses, chemicals, radiation) do so by altering the DNA apparatus within the cell so that it stops responding to the body's normal control systems.

Approximately 20 percent of cancers are caused by certain viruses, which are able to incorporate their own DNA into a cell for the purpose of making new viruses (opportunistic reproduction in its most manipulative form!). In the course of research on this came the discovery that cells have oncogenes (cancer-causing genes) within their normal DNA complement, which are usually repressed. Chromosomal abnormalities, like an interruption in the anaphase of mitosis or mutations caused by any number of things, can activate these oncogenes into causing cancer cells to proliferate.

There are two basic phases involved with cancer development. Stage one is the initiation phase characterized by DNA damage or alteration. DNA damage can be passed on to new cells in the form of mutations, but often the body can repair such damage or a healthy immune system can contain it. A mutated cell, referred to as neoplastic, is considered precancerous and does not become fully cancerous without the presence of a second factor, called a promoter. Precancerous cells can remain latent for decades and never become tumorous unless or until they come into contact with a promoter.

In view of the many studies pointing to EMFs at various frequencies as affecting alterations in calcium ion flow, mitosis, cell-growth enzymes (ODC), melatonin production, artificial stress response, and an increase in cell-receptor sites, as well as the simultaneous suppression of the immune system's natural killer T-lymphocytes, many researchers consider EMFs to be cancer promoters at the very least, if not outright initiators.

Cancer statistics appear to bear this out. According to one well-designed study published in the *Journal of the American Medical Association* (*JAMA*, Feb. 9, 1994, vol. 271, no. 6), the incidence of all cancers has steadily risen in the United States since the turn of the century, even when smoking and the growth of the population are factored in. The prevalence of petrochemicals throughout the globe is probably a factor, and so is the steady increase in the use of artificially produced EMFs in all frequencies across the spectrum.

Compared to what we know from the fossilized remains of Stone Age civilizations, the occurrence of breast cancer in women is estimated to be 100 times greater. Other studies have found that incidences of primary brain cancers and cancers of the endocrine system, of the blood (leukemia), and the skin (malignant melanoma) have risen sharply since the turn of the century, in some cases spiking 300 percent upward within the last two decades. A proliferation of microwave frequencies in particular is suspected.

Clearly something is turning our genetic apparatus on and off at cross-purposes with its normal sequence, activating the wrong things while suppressing the controls.

Some Electrotherapeutic Approaches

Cancer cells have unique electrical properties unlike those of other cells. Growing tissue is electrically negative, but cancer cells are the most negative in electrical charge of all the fast-growing cells. Cancer cells are not permanently stuck in their malignant state, as was once thought. "Once a cancer cell, always a cancer cell" is not true.

There is fascinating research into ways to convert cancer cells back to their former normal state. The approach is to "unstick" the cells from their incomplete dedifferentiated state and to dedifferentiate them the rest of the way, in the hope that normal growth will follow. It is probably the underlying biological mechanism behind the occasional spontaneous remission of tumors that so baffles oncologists.

Lab studies that have grafted cancerous tumors onto the regenerating limb sites of salamanders have shown those tumors to disappear completely in the process of regeneration. But, except for bone growth, humans do not regenerate body parts as some amphibians do. And those creatures that can regenerate limbs and other body parts rarely get cancer. In fact, the further away a species gets from the ability to regenerate, the more common cancer becomes. The salamander studies are promising, nevertheless. Something in the return to the dedifferentiation state overrides the genetic machinery of the tumor's oncogenes. This gives rise to the theory that the signal to dedifferentiate may be stronger than the oncogene signal to reproduce. If true, that is good news.

Unfortunately, research using various electrical polarities to treat cancer has produced alarmingly contradictory results. One of Dr. Becker's studies with human fibrosarcoma cells in cultures speeded up cell growth by over 300 percent through the use of both positive and negative current—much to the research team's surprise. (The team had anticipated that the application of positive current to the electrically negative cancer cells would produce a beneficial effect.) Other studies have made use of electrical properties in metals—silver ions in particular—combined with minute levels of positive current. This was found to halt mitosis in fibrosarcoma cells within one day and return them to complete dedifferentiation. Experimenting with metallic properties in conjunction with electrical currents might be a useful avenue of study, but caution is warranted, because many metallic ions are toxic.

Other researchers, noting the negative charge of cancer cells, have experimented with the introduction of positive current into tumors. Professor Björn Nordenström, a former head of radiology at the Karolinska Institute in Sweden, has done extensive research on the subject and has published a book on complex biologically closed electrical circuits. Dr. Nordenström inserted stainless-steel needle electrodes into inoperable lung tumors, with the visual aid of X-rays, and applied 10 volts of positive-charge electricity, combined with a negative electrode attached to the chest skin for the return loop. Most of his experiments have shown dramatic tumor reduction, but unfortunately nothing is as simple as it seems.

Dr. Nordenström, and others before him, was most likely killing tumors through electrolysis and tissue destruction called necrosis. Electrolysis is a process in which the water molecules within tissue are broken apart by the application of any voltage. This produces gases like hydrogen, which is highly toxic to cells. The higher the voltage, the more gases are produced. It has been found that voltage as low as 1.1 V will produce electrolysis in test animals.

So what's the difference *how* tumors are killed, just as long as they are? For one thing, not all of the cancer is killed. Surrounding tumors can grow wildly. And, since cancer cells are now known to proliferate rapidly with both positive and negative electricity, the likelihood that any such applications are causing distant tumors to grow cannot be denied, especially any cells along the return route from the negative electrode. Adjacent tumors may be stimulated rather than retarded.

In the second place, such electrical applications are not killing tumors through electrical processes alone or through another, larger biological control network within the body but, rather, as a by-product of chemical alterations resulting from electrolysis, localized toxicity, and necrosis.

Other researchers have used pulsed electromagnetic fields at various frequencies, with mixed results. The growth rate of tumor cells in cultures greatly increased in some studies. Studies in which the whole animal was exposed had contradictory findings: enhanced cancer growth in some, cancer reduction in others. Clearly we need to know a great deal more about the underlying mechanisms of biological interactions with EMFs before we try any of these approaches on the general population.

Some promising work has been done with steady-state DC magnetic fields in halting mitosis in cancer cells, but time-varying magnetic fields (what we are most exposed to in our ambient background) appear to accelerate cancer-cell growth.

Getting cancer cells to return to normal is a radically different approach from the one commonly used today, which is largely focused on killing the cells chemically or with radiation—the irony of these two approaches being that they are themselves carcinogenic. But a concerted attempt to get cancerous cells to fully dedifferentiate and return to normal is a promising area for cancer research. It unfortunately is not being explored with the kind of enthusiasm it warrants, despite the contradictory findings thus far.

Magnetic Materials in Body Tissues

Until just recently, it had been widely assumed that there was no inherent magnetic material manufactured within the human body. Then in 1992, Dr. Joseph Lynn Kirschvink, a professor of geobiology at the California Institute of Technology, Atsuko Kobayashi-Kirschvink, a research engineer also at Caltech, and Barbara Woodford, a research associate at the University of Southern California, found magnetic crystals in brain tissue in their studies of human biomagnetism sponsored by the National Institutes of Health.

In their studies, all areas of the brain revealed significant levels of magnetite, but the highest levels were associated with the meninges, the membrane that covers the brain. There were about

5 million crystals per gram of brain tissue, with concentrations of over 100 million crystals per gram in the meninges. Two types of magnetite were found, and much of it occurred in clumps. They were organized like cellular configurations called magnetosomes, the long crystal chains found in magnetic bacteria and fish, among other species, which use the earth's magnetic fields for vital directional information.

The kind of magnetite the researchers discovered (ferromagnetite) interacts a million times more strongly with external magnetic fields than with other biological material. It is not only permanently magnetic but is also capable of acting as a metallic conductor. The crystals that the Kirschvink research group isolated could be moved around by magnetic fields only slightly stronger than the earth's natural magnetic background.

This research suggests that if magnetite crystals are coupled with the cell's ion channels, external fields could be opening and closing these channels, with unknown biological effects. It is one plausible mechanism for the observed effects of the calcium efflux response at different frequencies.

The research may also point to the method by which we humans extract basic magnetic information from the world around us. In theory, cancer cells could be affected, too. In mice studies, tumors have proved to be a thousand times more magnetic than other body tissue. Also, the verification of magnetite in human tissue has implications for the increasingly widespread use of MRI as a diagnostic tool, calling its safety into question.

Similar work by Dr. Robin Baker, at the University of Manchester, England, found concentrations of magnetite and iron in the human ethmoid bone in the sinus region. This is the area of the body where some think our "magnetic organ" is located— high up in the posterior wall of the ethmoid sinus at the back of the nasal passage, just in front of the pituitary gland. The area is rich in nerve fibers extending along many paths into the brain.

Dr. Baker also conducted a series of experiments on the human ability to sense magnetic north, and he was able to disturb a test subject's sense of direction for up to two hours by applying a bar magnet to the forehead for only fifteen minutes.

The body contains other magnetic metals, such as iron and copper, which have slightly different magnetic properties than magnetite. What effect externally applied EMFs have on these metals is unknown.

Chapter 9

PREGNANCY, CHILDREN, AND
ELECTROMAGNETIC FIELDS

MANY PEOPLE think that a growing embryo, or a child after it is born, is physiologically just like a tiny adult. From a developmental point of view, nothing could be further from the truth. An embryo's growth into a child, then a teenager, and then a mature adult is marked by critical biological changes that adhere to each stage of one's development. Each stage harbors its own particular cellular activities that flower into the next.

A growing embryo or fetus, as well as all children, is in a higher state of mitosis, with greatly increased rates of cell division. The greater the cell division rate, the more opportunities there are for something to go wrong. Bioelectromagnetism plays a pivotal role in the early growth years, from conception through childhood.

There are only a few bona fide studies of electromagnetic fields and pregnancy in humans, and only a handful pertaining to children. But we can extrapolate certain conclusions from general cellular findings, from animal research (to some extent), and from epidemiological studies that have focused on pregnancy and children.

Defining Some Terms: Heat and Pregnancy

Three words often come up regarding adverse outcomes of pregnancy: mutagenic, teratogenic, and oncogenic or carcinogenic.

A mutagen can change the mother's or the father's reproductive cells, which can then affect their offspring through genetic damage. A teratogen can permanently alter fetal cells, causing a birth defect in the baby without necessarily harming the mother. Oncogens or carcinogens have the ability to make cells grow wildly. Some agents, like ionizing radiation, can do all three. In some animal studies, nonionizing radiation at some frequencies has been found to be teratogenic and oncogenic, and probably mutagenic, though it is unclear whether these observations were due to heating effects, nonthermal processes, or both.

Heat has great significance in pregnancy. Anything that raises the mother's core body temperature—like fever, vigorous exercise, or a long soak in a hot tub—to 102 degrees Fahrenheit (38.9 degrees Centigrade) can increase congenital anomalies, particularly during the first trimester, when all the fetus's major organs are forming.

A woman's basal body temperature is higher during pregnancy anyhow. It is possible for a woman working around radio or microwave frequencies (RF/MW), or living near certain broadcast towers, to increase her internal temperature to a dangerous level without even being aware of it. Remember that safety standards for occupational exposure, broadcast antennas, and consumer products were written for "the average male." The unique physiological processes of pregnancy and childhood were never factored in, other than to scale the power densities downward, on the presumption that women, growing fetal tissue, and children are just smaller versions of that average person.

Women dissipate body heat through the skin and through breathing. The growing fetus has an average temperature approximately half a centigrade higher than the mother and dissipates its heat across the placenta and through the mother's blood-stream. If either mother or baby cannot cool down quickly enough after a rise in the core temperature, fetal damage can occur.

The Occupational Safety and Health Administration (OSHA) has recommended limits for some workplace RF/MW exposures, but the limits are not mandated by law. No one knows how many women are being exposed to significantly higher dosages than the recommended ones—and those recommendations are for thermal effects only. Nonthermal effects could be causing problems ranging from early miscarriage and birth defects to learning disabilities and pediatric cancers (which were perhaps initiated prenatally).

The Period of Organogenesis

The first trimester of pregnancy is the most important one in many ways. By the end of the fourth month, nearly all of the major fetal organs, plus the brain stem and limbs, have formed and are maturing. Other than genetically inherited disorders, most pre-natal damage capable of causing birth defects occurs during this time. It is called the period of organogenesis or embryogenesis, and environmental insults of any kind can cause organ-specific damage to whatever is forming through cellular differentiation at a particular time.

Regarding any environmental insult, several factors come into play: dosage, length of exposure, and the specific time in fetal development. An embryo can react to a teratogen in one of three ways. A small dose may cause no effect; an intermediate dose may cause a pattern of organ-specific malformations; and a high dose may cause a miscarriage, leaving the actual cause of the damage unrecognizable to either the mother or her doctor. But in some animal studies with EMFs, the lowest dosages caused the worst malformations, so the relationship of birth defects and EMFs cannot be considered a linear cause-effect one, as with some chemical agents. A different process is involved with EMF-related exposure.

There may also be frequency-specific windows for different organ systems, meaning that certain exposures may be particularly detrimental at fixed times during fetal growth. This is well in advance of current reproductive research. However, given what we know about EMFs and cell differentiation, as well as the effect on mitosis, it is likely that complex links will eventually be discovered when more researchers take an active interest in the subject. The interactions will likely be both thermal and nonthermal. The notion of prenatal frequency windows could account for the sharp rise in learning disabilities and pediatric cancers, for instance, as well as for a range of seemingly unrelated birth defects that have been observed.

The fetal-growth timetable on the following page covers the period of organogenesis and beyond. Some organ development overlaps, and brain development occurs throughout the entire nine months, with many critical aspects late in pregnancy. In other words, there is no safe time for either strong intense bursts or low-level continuous doses of EMFs during the months of pregnancy.

How Some Frequencies Affect Pregnancy

EMF exposure during pregnancy affects both the mother and the fetus, and each has a different biological parameter to consider. Various frequencies are absorbed differently; some have only a superficial ability to penetrate tissue, while others penetrate deeply and can easily reach the fetal organs.

Two frequency ranges have received more attention than others. One is the microwave/shortwave range running from a few megahertz up to around 100 gigahertz, which is used for military communications. The other is the extremely low frequency (ELF) range of 10 to 60 hertz, which encompasses high-tension lines

Day 0	Ovulation
Day 5	Implantation
Days 19–27	Neural tube forms
Weeks 3–7	Heart forms
Weeks 4–10	Gastrointestinal tract forms
Weeks 5–8	Kidneys, limbs, and lungs form
Weeks 7–10	Genitals form; sex differentiation occurs
Weeks 10–16	Hair patterns form
Weeks 13–19	Skin organs, including sweat glands, hair follicles, and so on, differentiate
Month 4	Sensory and motor-nerve tissue matures
Month 5	Vaginal tube opens in females
Month 6	Lungs mature
Month 7	Eyelids separate

and most electrical equipment, including domestic wiring. Most of the research on both of these ranges has been thermally based, involving short exposures at high power densities, and has concentrated on the electrical component, not the magnetic one. It therefore gives us little useful information about long-term, low-level magnetic exposures, the kind we most often encounter.

The Federal Communications Commission (FCC) recently sold some of the higher gigahertz frequencies to large communications companies, to allow them to compete with cable TV, which functions in lower frequencies. New towers transmitting in the gigahertz ranges will likely be cropping up soon near residential neighborhoods and schools, along with cellular-phone towers transmitting in the upper microwave bands—all without the necessary biodata to predict health effects. Some companies plan to develop direct satellite-to-earth communications systems, in effect blanketing the earth with a new range of frequencies, but at least the ground-level exposure will be much less powerful. While satellite broadcasting will probably prove to be safer than ground-level broadcasting, even this is unclear with respect to reproductive outcomes.

There is no question that the thermal effects of RF/MF radiation pose a significant threat to pregnancy. Radiation is absorbed in the "average" person most efficiently and deeply around 77–87 megahertz, but keep in mind that this hypothetical person is not necessarily female and certainly not a fetus. It is thought that significant heating does not occur at the transmission powers allowed by the FCC. But some radar frequencies in the microwave bands are capable of inducing thermal and subthermal biological effects in humans and in test animals, and heating can occur at any frequency if the wave is strong enough. Diathermy equipment, for instance (which operates at 27.5 megahertz), has great penetration ability and can cause significant internal heating, and therefore fetal damage.

If there is a possibility that you may be pregnant, do not allow diathermy equipment to be used on you, especially in your abdominal area. You may be in the early stages of pregnancy without knowing it if you are not using birth control methods and are having sex regularly.

By way of contrast, many workplace industrial exposures are in the 10-to-40-megahertz range, and a microwave oven operates at around 2,450 megahertz. It is thought that microwave frequencies above 10,000 megahertz have little ability to penetrate, but this does not mean that complex biological actions are not being set into motion.

Hazardous ELF/RF/MW work environments for both men and women of reproductive age are those near antenna transmitters of any kind, generator tubes, or other high-frequency equipment, as well as at electrical substations. Other occupational hazards would include heat-sealing equipment, radar equipment, and strong automatic-security systems. If a simple radar detector in your car beeps constantly near your place of employment, it could mean that the workplace is blanketed with radar frequencies. The home environment would also be considered hazardous if it is near such facilities.

Animal research has repeatedly shown adverse reproductive effects from various RF/EMF exposures, and frequency windows are commonly found. Rats exposed to 35 milliwatts per square centimeter (35 mW/cm^2) of continuous microwaves in the gigahertz range on days 6 to 15 of gestation showed an increase in fetal malformations, but no effects were seen in mice exposed to power densities of 5 to 21 mW/cm^2. In other studies, rats exposed to 915 megahertz at a power level of 10 mW/cm^2 (within the cellular-phone frequency) were found to have no significant maternal heating.

Some researchers have deduced that the 30 mW/cm^2 range is the most teratogenic to rodents, and they have extrapolated from this that the ANSI standard of 1 mW/cm^2 for frequencies of between 30 and 300 megahertz provides adequate protection for pregnant women and their embryos/fetuses. This kind of extrapolation raises several problems, the most glaring being that rodents have a totally different anatomical structure and shape than humans and therefore conduct electromagnetic fields quite differently. This is in addition to the differences in size, tissue construction, and a host of other physiological factors between the species.

Tissue response to EMFs varies greatly according to the tissue's water content. Skin and muscles are more sensitive than bone, for instance, and organs like the bladder, the testicles, and the brain are more sensitive still. But no one has researched the possible resonant frequencies of embryonic fluid. Perhaps such a line of inquiry could yield clues to the miscarriage clusters that occur in pregnant computer-terminal operators. Efforts so far have centered on a possible direct impact on the fetal tissue, not on resonant changes in the fetal environment.

Thermal receptors in the mother's skin are generally insensitive to frequencies below 1,000 megahertz, which are easily absorbed by the body, and women are not especially sensitive to temperature changes within the uterus. Thermal reactions could be occurring without the mother knowing it—not to mention nonthermal effects.

Nonthermal Effects

Because of the scarcity of research on pregnancy and ELF/RF/MW, it is difficult to know when a problem is thermally caused and when it is due to more subtle nonthermal interactions. Of course to the couple experiencing the trauma of miscarriage or birth defects, the distinction is purely academic, but it is still an important issue for researchers.

One interesting body of work, however, by epidemiologist Nancy Wertheimer and physicist Ed Leeper, concerns both thermal and nonthermal EMF effects on pregnancy and on cancer in children. Over a number of years during the 1980s and 1990s, the researchers found as much as a 50 percent increase in miscarriages in women who slept in electrically heated waterbeds, under electric blankets, or in rooms with ceiling-cable heat (which is like installing a huge electric blanket over a contained space). The miscarriage rate rose dramatically during the winter months between September and January, when settings on electric blankets are consistently higher.

Electric blankets give off relatively high magnetic fields, mostly because of the way the wires are positioned in relation to each other. In the middle of the blanket, the current flows in an S pattern of parallel wires, which would normally cause the fields to cancel each other out, but near the edges the current is unbalanced. (This is especially true of blankets manufactured prior to 1992.) For any heat setting between one and ten, Wertheimer and Leeper measured the fields at between 10 and 20 milligauss right next to the blanket, which would be directly on top of the sleeper for eight hours, and at as much as 5 to 10 milligauss six inches away. (Blankets with separate controls for the two sides of the bed would be subject to the highest heat setting for the whole blanket.) Their measurements for electrically heated waterbeds were slightly less, at around 5 milligauss, since the heating element on most models is on the underside of the bed. That is nevertheless a high exposure for such a long duration.

These measurements are especially worrisome in light of studies that have found childhood cancers in homes with ambient background exposures of only 1 to 2 milligauss, as well as numerous animal studies showing severe birth defects at lower exposures. Wertheimer and Leeper suspect that the incidence of miscarriage may in fact be much higher than their surveys indicated, due to the fact that many early miscarriages go unnoticed since women are typically 2 to 4 weeks pregnant before they even miss a menstrual period. Women experiencing early miscarriage often think their periods are just late.

Of course with electric blankets and heated waterbeds, thermal damage can easily be imagined, but fetal abnormalities have consistently been observed from EMF exposures that are not due to heat alone. Similar findings of increased fetal loss were noted in homes with ceiling-cable heat, which raised the ambient exposure to 10 milligauss. Also, increases occurred during the cold winter months, when heat use (and therefore magnetic-field exposure) was greatest. But since the cable heat is a number of feet overhead, thermal-effects-only could be reasonably ruled out. Families living in homes without ceiling-cable heat had normal miscarriage rates.

In response to safety concerns, some electric-blanket manufacturers have redesigned their newer models to produce lower fields. Electric blankets manufactured prior to 1992 should be discarded. The best advice if you are planning pregnancy, are already pregnant, or have had miscarriages is to heat the bed before you get in—then completely unplug the blanket (don't just turn it off) since voltage is present in the wires whether the blanket is turned on or not. If you sleep in a heated waterbed, try to change to a regular bed until after delivery. Don't use an electric heating pad, and if you live in a home with ceiling-cable heat, try to replace it with some other kind until after delivery.

Additional Studies and Adverse Outcomes

In the power-line frequencies, several animal studies have found disturbing reproductive outcomes. We noted in a previous chapter the work of neurophysiologist José Delgado, formerly of Yale University and later at Centro Ramon y Cayal in Madrid, Spain. Dr. Delgado found gross deformities in chick embryos exposed to ELF magnetic fields at 10, 100, and 1,000 hertz. Most of the major developmental defects were found at the 100-hertz level, in strengths as low as 1 milligauss. In 1986, the Office of Naval Research backed an international study at six different laboratories, five of which confirmed Dr. Delgado's findings that very low-level, very low-frequency pulsed magnetic fields contributed to abnormal development in early chick embryos.

There is also the work that Drs. Andrew Marino and Robert Becker conducted on three generations of rats. They found severely stunted growth (especially among males); large increases in infant mortality (rats born during various generations failed to live to maturity at a rate between 6 and 16 percent higher than the normal death rate); and an increased incidence of weight gain due to water retention, as well as underweight test animals—both effects attributed to abnormal stress responses. Dr. Marino's ten original experiments were conducted for one month at 60-hertz electric fields of 100 to 150 volts per centimeter—the equivalent of the typical exposure under a high-tension line.

A number of studies by various agencies between 1976 and the mid-1980s tried to replicate these studies, with results that were contradictory, controversial, and disturbing. The federal

Energy Research and Development Administration (ERDA) in 1976 commissioned the Battelle Pacific Northwest Laboratories in Richland, Washington, to replicate them, as did the Electric Power Research Institute (EPRI). The Department of Energy also commissioned Battelle to make a three-generation study of mice exposed to 60-hertz fields.

Battelle, then headed by Richard D. Phillips, came up with findings that were not uniform across all the experiments, but it, too, found severe growth retardation over three generations of test animals, as well as much greater weight variations than normal. In separate sets of tests using miniature pigs (chosen because their body weight approximates that of the average human), as well as rats, results included melatonin and endocrine-gland suppression and a threefold increase in birth defects among the offspring of female rats chronically exposed to 60-hertz fields.

In the mini-pig studies, almost all the exposed animals died in an epidemic, but far fewer of the nonexposed animals died. (Chronic stress caused by the fields is suspected to have lowered the animals' immune resistance.) As the studies continued, it was found that first- and second-generation offspring of exposed females that were born and bred in 60-hertz electric fields (at 30,000 volts per meter) but were mated with unexposed males were significantly lower in body weight and had twice as many birth defects.

Several of the commissioning agencies, as well as the final Battelle reports, issued statements saying that no evidence of harm was found in some studies, and that the overall results were inconclusive. Phillips left Battelle and went to the EPA's Experimental Biology Division at Research Triangle Park in North Carolina. He later said that contrary to those official statements, they had in fact found a marked reduction in nighttime pineal-gland melatonin production; significant reduction in testosterone in male rats; alterations in neuromuscular systems; and serious increases in birth defects in both rats and pigs chronically exposed over two generations.

Some studies have been done on human genetic defects and EMFs. In 1983, Dr. S. Nordstrom and a research team at the University of Umea, Sweden, found a significantly higher number of malformed children among children whose fathers worked in electric switchyards and were exposed to high voltage fields. Dr. I. Nordenson (a colleague of Nordstrom's) found a significant increase in chromosomal abnormalities overall in the white blood cells of similar workers. These studies indicate that the 50-hertz fields used in Europe cause genetic damage to the sperm cells of the fathers, which is passed on to their children in the form of birth defects—in other words, ELFs/EMFs are mutagenic.

Similar observations were made by Drs. E. Manikowska-Czerska, P. Czerska, and W. Leach at the Food and Drug Administration's Center for Devices and Radiological Health, in 1983, in studies exposing male mice to microwaves. They found reduced sperm production after short nonthermal exposures of only thirty minutes per day for two weeks, as well as significant chromosomal abnormalities in the sperm. When exposed male test animals were mated with unexposed females, far more miscarriages occurred. The researchers concluded that microwaves in dosages far below the level of thermal exposure caused genetic damage directly to parental chromosomes, which could be passed on to offspring. But such effects can also occur directly in a growing fetus in utero.

There is also a link between fathers who were military radar operators and an increase in Down's syndrome in their children, as reported by Dr. A. T. Sigler in 1965 in the *Bulletin of the Johns Hopkins Hospital*. Later studies that failed to confirm this work were considered suspect by critics, but no new studies on Down's have been done since. Some researchers think that a much higher incidence of Down's syndrome (nearly 1,000 percent above the national average) and other birth defects around Vernon, New Jersey, where satellite uplinks in the microwave bands have proliferated, is attributable to the constant exposure of the residents.

Dr. Reba Goodman, at Columbia University, is finding

complex relationships with different results that are both frequency dependent and vary according to the cell type being exposed. Her ongoing work is directed at the chromosomal effects of ELF/RF exposures on both humans and insects.

EMFs and Infertility

Although few studies have been done, some now think that the steep rise in infertility over the past decade, especially those cases for which no cause can be found, may eventually be attributed to EMF exposure. A full 10 to 18 percent of Americans in their reproductive years may be infertile at any given time. This coincides with a sharp decline in the general sperm count among American men, to nearly half that of the 1950s.

In women, infertility can be due to menstrual-cycle irregularities, hormonal imbalances, and disorders of the ovaries, fallopian tubes, uterus, or cervical mucus, as well as to DES exposure and chronic miscarriage. In men, infertility can be caused by abnormal sperm production, sperm damage, blockage of the sperm flow, and various medications, illnesses, and chemical exposures. Fertility declines gradually in both men and women after the age of thirty.

Sperm production is affected by heat. The testicles must be slightly below the normal body temperature in order for sperm to develop. If the genital temperature is elevated for any length of time, sperm production is diminished. Research with Navy personnel dating back to the 1940s has found lowered sperm counts and sperm damage in radar operators. Truck drivers with CB radios positioned close to the genital area could have an elevated testicular temperature, thereby doing sperm damage. The same would be true of hand-held radar guns kept near the genital area—a common practice among law-enforcement officers—and it remains to be seen what effect car phones with a power supply positioned at genital height may have. Men who work in overly hot environments are also known to have lower sperm counts.

Nonthermal effects on fertility have been noted as well. Testicular degeneration has been clearly demonstrated in animal studies in the RF/MW ranges. As noted in many of the studies already cited, impaired fertility has been repeatedly found in exposed animals in the form of reduced pregnancy rates and smaller litters. Pregnant females have been found to have fewer uterine implantation sites, increased malformations in offspring, and higher miscarriage rates.

Electromagnetic fields may also be having an impact on fertility in two roundabout ways. In Chapter 7, we explored the relationship of EMFs to melatonin and hormonal balance, and the ability of EMFs to create artificial stress conditions that could be measured chemically in people and test animals that did not appear especially anxious. Stress is known to affect fertility through the hormonal levels of both men and women.

The body responds to stress by shutting down the oxygen and blood supply to peripheral body systems and concentrates it in vital fight-or-flight organs like the heart and brain. The testicles need a substantial amount of oxygen to produce sperm. Anything that depletes oxygen to the testicles can adversely affect sperm count. In women, stress upsets the entire hormonal axis through the pituitary and the hypothalamus, which in turn controls ovulation and several other subtle but essential phases of reproductive ability. Conceiving a child requires more fine-tuning than many people realize. In any given monthly cycle, even couples with unimpaired fertility have only a 25 percent chance of conceiving under the best of circumstances. Anything that marginally alters those odds can tip the balance in the wrong direction for some couples.

There are today many unexplained cases of subfertility, as well as a range of odd immunological reactions, such as a woman's developing antibodies to her mate's sperm or to the embryo itself, which make conception very difficult at best or carrying a child to term nearly impossible. Given some of the research on EMFs and their effects on the immune system, connections may eventually be found there, too, but this is purely speculative at this time.

For anyone experiencing impaired fertility, common sense would dictate a reduction in exposure to EMFs at work, at home, and in the use of various EMF/RF-emitting consumer products.

Miscarriage Clusters and VDTs

Eleven birth defect and miscarriage clusters have been noted in the United States and Canada (with additional occurrences throughout Europe and in Australia) among pregnant women who worked at CRT (cathode-ray tube) computer-display terminals. A cluster is an above-average episode in one location that is statistically higher than what would normally be expected. Typically, the woman relates her health problems to her doctor and her employer; when enough complaints accumulate, the employer will call the manufacturer of the computer equipment, and a representative of the company will investigate. Sometimes field measurements are taken, sometimes not. Sometimes health and labor organizations become involved, sometimes not.

With the earliest clusters, field measurements were invariably found to be within safe limits, and the machines were deemed harmless. It later turned out that the measuring devices were inadequate and often did not measure the lowest frequencies, which were the most abundant. Today, although there is still much official stonewalling, most computer manufacturers have developed low-emission models and are voluntarily adhering to Sweden's more stringent safety standards—at least a step in the right direction. Many older models (1988 and older) can give off particularly high fields; those manufactured before 1982 should not be used.

Work-station design recommendations will be made in Chapter 13, but if you are planning pregnancy or are already pregnant, here are some important things to know right now.

In 1988, Drs. Marilyn Goldhaber, Michael Polen, and Robert Hiat, at the Kaiser Permanente Health Group in Oakland, California, conducted a well-designed study of 1,583 pregnant women. They found that those using computers for more than twenty hours per week had twice the miscarriage rate (an incidence of over 40 percent) of females doing similar office work without computers. (A subsequent study by the National Institute for Occupational Safety and Health did not replicate these findings.) Recent Finnish studies confirmed a fourfold increase in miscarriages in women using computers that produced unusually high levels of extremely low frequency magnetic fields.

Other researchers have reported a number of adverse medical conditions associated with CRT computer use. Along with the increased miscarriage incidence, these include birth defects in babies born to computer operators, cataracts, various vision difficulties, disturbances in the menstrual cycles, skin rashes, chronic stress, headache, nausea, sleeplessness, and fatigue.

Animal studies using the computer frequencies have also found patterns of adverse reproductive outcomes. Several Swedish research teams found five times the incidence of developmental abnormalities in mice exposed to the common computer frequencies over unexposed control animals. And researchers at Purdue University found that they not only could produce abnormalities in chick embryos, but that the defective genes afterward carried a reduced electrical field. This suggests that natural electrical fields play a crucial role in directing embryonic development, just as Dr. Robert Becker has postulated.

With so many women of reproductive age using computers, some precautions to minimize the risk during pregnancy are in order. Sit at least thirty inches back from the screen itself if you can. Usually this can be done if your system has a detachable keyboard. Once pregnancy is confirmed, or if you have already had miscarriages or children with birth defects or learning disabilities, request noncomputer work instead. At the very least, limit your computer use to no more than twenty hours per week—including any personal computer work done at home.

If you are reassigned to a noncomputer job, be sure your new work location is not right next to someone else's computer. The backs and sides of the machines typically have the strongest emissions. Magnetic fields penetrate through walls, so investigate the positioning of equipment in adjacent offices. If possible, get an inexpensive gaussmeter and make actual measurements in your environments. (You're looking for measurements no higher than 2 milligauss, preferably lower than 1 milligauss.) Other office machines can produce electromagnetic fields as high as or higher than computers. And how various office machines are positioned in relation to each other can count, too. (See Chapter 13 for more information.)

Diagnostic Tests

Ultrasound

Prenatal frequency windows could also have profound implications regarding the use of ultrasound equipment routinely applied during the course of a pregnancy in today's obstetrical practices. Some now think it should be used far more judiciously than it has been in the past. Unfortunately this new less-is-more idea is based on financial considerations and the fact that the use of ultrasound has not changed adverse obstetrical outcomes anyhow. It is not yet based on the growing suspicion that ultrasound may not be as safe as once thought. Curiosity about the baby should never outweigh the risks involved with ultrasound. It is not a completely benign viewing tool like many women (and health practitioners) have come to believe.

Although still largely considered safe, ultrasound devices use strong RF sound frequencies with deep penetration ability, in dosages set more with the mother's anatomy in mind than that of the growing fetus. Ultrasound can create temperature elevations in both maternal and fetal tissue, as well as a process called "cavitation"—the creation of resonance within cells, which can cause a host of adverse fetal reactions. Both have been observed with certain ultrasound applications—although mostly at higher levels than those used in standard prenatal tests.

The fact remains that the casual repeated use of ultrasound is not a good idea. Ultrasound equipment is now common, for instance, in veterinarians' offices. Some pregnant women working in such offices presume that ultrasound is safe enough to "look at their babies" anytime. Nothing could be farther from the truth, and this is not recommended.

Exposures to nonionizing radiation are cumulative, just as they are with ionizing radiation. All exposures, including those from other ultrasound applications, like echocardiography, should be kept to a minimum, beginning prenatally.

With any prenatal test, the risk/benefit ratio comes into play. Are you using ultrasound to see if the baby is normal? If the baby is not normal, would you terminate the pregnancy? Or are you opposed to abortion under any circumstances? If so, then the only benefit of ultrasound would be the added time in which to prepare for a child with special needs, a benefit that might not outweigh the risk to the fetus. Along another line, do multiple births run in your family? Are you taking fertility drugs that could increase the likelihood of a multiple pregnancy? Such pregnancies have a range of risks to mothers and babies, so the benefit of that advance knowledge would perhaps outweigh the risks from the ultrasound. Each case is different. The point is, it's time to start thinking of ultrasound in this way.

Magnetic Resonance Imaging

Magnetic resonance imaging (MRI) is a diagnostic tool being used instead of X-ray to find abnormalities in soft tissue and glands. It subjects the body to static magnetic fields in excess of 20,000 times the earth's natural background, as well as to strong

RF frequencies. It is touted by many diagnosticians as perfectly safe because, unlike X-rays, it uses nonionizing radiation. That opinion may prove to be naive at best, and criminal at worst. Some researchers are beginning to express reservations, especially with much stronger MRI units making their way into some hospitals.

MRI works by using a powerful magnetic field to align the hydrogen molecules of the body at the cellular level. Then a strong radio-frequency pulsed wave is introduced, and the magnetic field is cut off. As the hydrogen atoms revert back to their normal position, they give off intense bursts of electromagnetic information, which is recorded as an image. One problem (among many) is that no one knows just what that energy burst is that the body is producing, and the presumption that the molecules go back to normal may prove to be wishful thinking.

MRIs create three disparate electromagnetic fields in an exotic combination: an unchanging (static) magnetic field; another kind of magnetic field called a gradient field, which increases in intensity across the body and provides the scanner with information on where the tissue being imaged is located; and rapidly pulsed radio-frequency energy. Most MRIs today use magnetic fields of between 0.12 and 2.0 teslas (1 tesla equals 10,000 gauss and 10,000,000 milligauss). The new machines propose using four to ten times that strength, as well as stronger radio frequencies and more rapidly switching gradients. Some machines are planned for use on the body's sodium and lithium molecules, too.

In 1991, the first (and to date only) conference to investigate the safety of MRI was held by the New York Academy of Sciences—several years after the machines had been used on thousands of people, supposedly without adverse effects. Those present repeatedly spoke of the lack of information on the bioeffects of such exposure to humans. The early safety studies used machines of lower strengths than the ones now used on people, plus only a handful of studies were done, on cell cultures or animals, before the machines were allowed access to the marketplace. When MRIs made their debut in diagnostic centers, radiologists were flabbergasted at the clarity of their images, and doctors (by far the majority of MRI investors) embraced the technology as safer than X-ray, without knowing very much about electromagnetic fields on the lower end of the spectrum.

Anyone who has had an MRI is an unwitting test subject. MRI proponents say that if the technology was unsafe, this would have been immediately apparent, that lots of people have had them without problems. But no one is actually keeping tabs. There are several flaws in that logic — many adverse effects have a long lead time, but, more importantly, there are no channels for that adverse information to travel. This isn't a trickle-up situation like many of us presume, and people have experienced all sorts of odd things in MRI chambers that our normal vocabulary has trouble expressing.

Approximately 20 percent of those having MRIs experience severe panic; some lose consciousness for short periods of time; some undergo a kind of physical dissociation of mind from body; many report a sense of "losing it" or feeling "weird" in a way they cannot describe. A large percentage have to be sedated.

MRI proponents attribute the panic attacks to claustrophobia because of the confined space of the MRI "doughnut." But many of those who panic have no previous history of claustrophobia, and similar doughnut chambers, like CAT scans, do not cause the same high degree of panic. The other reactions are dismissed as due to the noise of the MRI (a loud knocking sound) or the length of time needed for the test (usually about forty-five minutes). But those who have had such adverse reactions insist that something else is occurring that has nothing to do with noise or duration — they just don't know what.

More tangible physical reactions to MRIs include short-term memory damage, thought to be reversible but lasting in some people for indefinite periods; localized tissue heating, which is thought not to exceed 2 degrees Centigrade but may in fact be much higher in areas of the body that do not dissipate heat easily (the eyes, testes, brain); a sharp metallic taste in the mouth lasting several days, thought to be due to induced electrical currents across the tongue, but that may have something to do with loosening the fillings in the teeth; and such symptoms as tinnitus, vertigo, headaches, nausea, muscle twitching, abnormal cardiac rhythms, increases in blood pressure, and a kind of visual sparkling-light phenomenon called magnetophosgenes, which no one understands very well. Some have experienced an unexplained skin burn and knocking or shocklike sensations.

Like many other studies of animals exposed to various EMF frequencies, early animal MRI studies found an increase in natural killer-cell cytotoxicity, indicating an initial stimulation of the immune system; alterations in the blood-brain barrier; and a decrease in nighttime drug absorption rates that indicate MRI's ability to interfere with calcium binding in brain cells, as well as changes in pineal-gland activity and melatonin production. These animal studies follow the same general research patterns revealing how EMFs affect the anatomy.

The normal risk/benefit equation may be lopsided with regard to MRI, that is, the risk may be significantly higher than is generally understood so it's more difficult to accurately weigh it against the benefit. MRI should be used only after other noninvasive tests have been tried, not as a first choice. And MRI is definitely not recommended during pregnancy, and probably should be avoided by anyone actively trying to become pregnant. If you are disabled or ill enough to require an MRI, perhaps pregnancy is best delayed for a while anyhow.

Even the areas around MRI test chambers have powerfully elevated fields, and some concern has focused on the MRI technicians who are exposed to them, even though they spend most of their time in a console room rather than close to the magnetic core of the unit. Nevertheless, one recent study found a threefold increase in the incidence of miscarriage in MRI workers as compared with homemakers. The same study, done by Drs. Josephine Evans, David Savitz, Emanuel Kanal, and Joseph Gillen, did not find an increase in infertility or low birth weight in infants born to MRI operators, however.

Recent animal research done by Drs. Meike Mevissen, Siegfried Buntenkötter, and Wolfgang Löscher at the School of Veterinary Medicine in Hannover, Germany, found increases in fetal resorptions—similar to miscarriages in humans—in pregnant rats exposed to field levels comparable to those of MRI technicians. Minor skeletal anomalies were also significantly increased in the offspring of test animals.

As a precaution, female MRI technicians who are trying to conceive should consider changing jobs until after delivery, especially if they have experienced prior miscarriages. Although the recommendation is purely speculative at this time, male MRI technicians who are experiencing impaired fertility may also want to request different work at least three months before trying to conceive—the length of time it can take to manufacture certain sperm cells.

For additional information on occupational exposure, see Chapter 13.

Electrotherapeutic Devices

There are a host of electrotherapeutic devices that employ RF/MW frequencies or low-level electrical stimulation of nerve tissue, mainly to relieve pain. Diathermy uses RF/MW to deep-heat localized tissue, much the way a microwave oven heats food. Diathermy is not recommended during pregnancy, especially near the abdominal area, and anyone trying to become pregnant might want to forgo its application as well. (For additional information on occupational diathermy exposures, see Chapter 13.)

Another kind of electrical device in widespread use today is called TENS, which stands for transcutaneous electrical nerve stimulation. There are many different types with other acronyms (like MENS, for muscle electrical nerve stimulation), but they are all essentially TENS units.

Originally such devices were developed to relieve chronic pain, after it was observed that differing amounts of pulsed electrical current at various frequencies could do just that. But the underlying physical explanation remains unknown. Unfortunately (due to a loophole in FDA regulations), the use of TENS-type devices has expanded to include many other clinical applications, including treatment of migraine, drug addiction, memory loss, and bladder dysfunction; wound and bone healing; reduction of tissue swelling (edema); the dissolving of some blood clots (hematomas); and as an electrical stimulator in acupuncture treatments. None have any grounding in scientific theory.

Dr. Robert Becker, in his *Cross Currents,* recommends that the use of TENS-type devices be limited to cases of chronic pain and that their applications be restricted to areas of the body other than the head, neck, and spinal cord. Anyone with a history of cancer should avoid them altogether, given the conflicting research on electromagnetic fields as both stimulators and suppressors of tumor-cell growth. TENS devices should be avoided during pregnancy, as should any device that creates a high electromagnetic field in close proximity to the mother (and therefore the baby). That includes electric-massage chairs or tables, heating pads, and the like.

Children

Childhood is a time of great cell growth, as we have noted, so there are particular concerns regarding children's exposure to EMFs. This includes everything from living near high-tension lines, to their use of computers at home and at school, to sitting too close to the television set or using a portable phone.

Researchers Nancy Wertheimer and Ed Leeper, in some of their earliest work in the Denver, Colorado, area, came up with some startling findings. After several years of painstaking field study, they discovered that the 60-hertz magnetic field component in strengths of only 3 milligauss was responsible for a two-to threefold increase in childhood cancers—leukemia and brain tumors in particular. The increased cancer rate occurred in the first and second houses nearest to step-down transformers (large cylindrical barrels mounted on poles) drawing their electrical current from standard neighborhood wires. What was so surprising was that these strengths are lower than the earth's natural fields, and certainly lower than those at the edge of high-tension-line rights-of-way.

Wertheimer and Leeper's work was repeated by Dr. David Savitz, from the University of North Carolina, who obtained the same results, reporting that 20 percent of childhood cancers were attributable to exposure to 3-milligauss, 60-hertz magnetic fields. Similar studies have been reported in Denmark, where researchers found a significantly increased risk of childhood lymphoma in homes closest to high-tension lines. However, the incidence of childhood leukemia and brain tumors was not significantly elevated in the Denmark studies.

Learning Disabilities

There has been a staggering increase in learning disabilities over the last two decades, including attention-deficit disorders, complex forms of dyslexia, hyperactivity, and minimal brain dysfunction, to name but a few. While learning disabilities can be caused by any number of physical or psychological factors—and statistical spikes resulting from better identification and reporting may account for part of the rise—there is also a possible link with electromagnetic fields that warrants more investigation.

Animal studies have linked 60-hertz fields with an inability of test animals to perform simple tasks. Dr. Kurt Salzinger, of the Polytechnic University of Brooklyn, exposed rats to 60-hertz fields during fetal development and the first few days of life, after which they were raised normally until they were three months old. They were then trained, along with unexposed rats, in different learning routines. The exposed rats were found to make many more mistakes and to learn much more slowly than their nonexposed counterparts. Not only were the differences unmistakable, but they continued long after the test animals' exposure. (No one has yet tried to replicate this important work.)

Dr. Frank Sulzman, of the State University of New York, also conducted 60-hertz studies on the biological cycles of monkeys, in which he found a significant reduction in lever-pressing activity to obtain food. This lowered activity lasted for months after the exposure. And Dr. Jonathan Wolpaw, of the New York State Department of Health, investigated neurohormone levels in the spinal fluid of monkeys exposed to 60-hertz fields for three weeks. He found significantly depressed levels of seratonin remaining below normal levels for months. (Lowered seratonin in humans has been implicated in suicide, depression, and behavioral abnormalities.)

While the low dopamine levels that Wolpaw reported returned to normal shortly after the exposures stopped, the suppression of dopamine is in itself interesting in relation to Parkinson's disease. Parkinson's, a degenerative disorder of the central nervous system caused by a lack of dopamine production in the brain, was until recently considered a disease exclusive to the elderly. But patients as young as eleven years old are now being diagnosed with it, sometimes after several years of going from one doctor to another simply because physicians never associate the disorder with anyone under fifty-five years of age.

Not only is the total incidence of Parkinson's rising in the global population, but one estimate holds as much as a 50 percent rise in cases in people under the age of fifty. An environmental stimulus of some kind is suspected. A link between EMFs and Parkinson's is purely speculative at this time, but it may be fertile ground for research since no one has any idea why it is occurring so much more frequently and in younger people for the first time.

Other disorders diagnosed in childhood that may be associated with prenatal EMF exposure include autism and a genetic disorder called Fragile X syndrome, second only to Down's syndrome in occurrence in the general population. Nearly identical brain lesions of the cerebellum have been found in a majority of autistic children that coincide with lesions created in test animals exposed to 50-hertz power-frequency fields and low-power microwaves.

Fragile X syndrome is a sex-linked genetic disorder marked by a weak or separated fragment of one of the limbs of the X chromosome. (X and Y chromosomes determine whether we are male or female: XX for female, XY for male.) The syndrome can be very mild and hardly recognizable, or it can include mental retardation, behavioral problems, and certain facial characteristics. Brain lesions deep in the cerebellum that are very similar to those seen in autistic children and EMF test animals are also found in those with Fragile X.

Sudden Infant Death Syndrome

There may also be a link between EMF exposure and the tragic occurrence known as sudden infant death syndrome (SIDS), in which an apparently healthy infant dies during sleep for no apparent reason. (It is also called crib or cot death.) Many pediatricians have the impression that it has increased in incidence in recent years.

Dr. William Sturner, the chief medical examiner for the state of Rhode Island, has found a link between SIDS infants and low melatonin levels in their brains, as compared with infants who died of known causes. (The melatonin levels in SIDS infants averaged 15 picograms per milliliter, compared with 51 picograms per milliliter in the control group.) The low level of melatonin may have led to enough of a decrease in respiratory function to cause infants to stop breathing.

Although no causal link is yet established, given the numerous studies now pointing to melatonin suppression and EMFs, such a link would not be surprising. In fact, Dr. Cornelia O'Leary, of the Royal College of Surgeons in England, reported that eight SIDS deaths occurred over one weekend, four within the same two-hour time span. It turned out that all the deaths were within a seven-mile radius of a military base testing a powerful new radar system. Dr. O'Leary's work on SIDS and electromagnetic fields is ongoing.

Conclusion

A broad and disturbing picture is beginning to emerge regarding electromagnetic fields and their effect on growing organisms in general, including the human embryo/fetus and children. Two specific parts of the body appear to be primarily affected: the brain and growing tissue (which would include fetal tissue, as well as rapidly dividing cancerous tissue).

Adverse effects to the brain appear to be primarily functional, resulting in learning disabilities, behavioral dysfunctions, and changes in hormone levels and circadian rhythms. Effects on growing tissue include heightened abnormal cancerous growth, an increase in birth defects, and interference with the genetic apparatus.

As consumers living in a time of emerging scientific consensus and an accompanying regulatory vacuum, it is up to us to assume the responsibility for reducing our EMF exposures. We can only hope that government regulatory agencies will soon assume their proper responsibility for protecting our future generations. But for now, it's up to us.

Chapter 10

TWENTIETH-CENTURY MALADIES

WE MAY BE dealing with new diseases that humankind has never seen before; increases in disease states that have plagued us for years; and changes in existing diseases that medical science thought it had learned to control. When new diseases seem to come from nowhere, there are two theoretical ways this occurs. One is through a genetic alteration in a bacteria or virus, which effectively creates a new microorganism. Another is a change in resistance to disease in general, making people (and other living creatures) more susceptible to illnesses. We are probably seeing a combination of both these avenues, and EMFs may play a pivotal role.

Many of the "new" diseases we are seeing, like AIDS and chronic fatigue immune dysfunction syndrome, fall in the category of immune disorders. (Immunology itself is a relatively new area of medical concentration over the last ten to fifteen years.) Cancer is thought to be not only pathological cell growth but also an inability of the immune system to keep abnormal cells in check.

Changes in existing diseases are also occurring. Learning disabilities and developmental defects in children have increased tremendously, as have Parkinson's and Alzheimer's disease. Are we seeing the effects of our completely altered electromagnetic environment come to fruition? Although still speculative, many studies seem to point in that direction.

Immune-System Disorders

There may be indications that EMFs influence several immune-system disorders in humans, and possibly in animals, too. One thing is for sure: immune-system disorders are on the rise, including allergies in general, autoimmune diseases in particular, and the appearance of new diseases like AIDS and AIDS-like illnesses now afflicting not only humans but sea mammals such as dolphins and seals as well. Several forms of cancer, especially of the skin, brain, breast, blood, and lymphatic system, are all on the rise since the early 1970s, in percentages that exceed general increases in the population and that cannot be accounted for numerically with earlier detection methods. In addition, EMFs are known to cause genetic mutations in bacteria and viruses, and several new strains of each are puzzling researchers. New strains of antibiotic-resistant bacteria have become life threatening all over again. Viral outbreaks in Cuba and in the American Southwest are reminiscent of Legionnaires' disease, although with different strains of virus acting in new ways. Compromised immunity plays a role in all of these things.

For many years, the prevailing view of immunologists about the role of the immune system has been to recognize "self" from "non-self." But a new model is emerging. The immune system may be more involved with recognizing danger—in the form of anything that causes tissue stress or destruction—than the former view. Some enigmatic immune-system disorders make more sense according to this new model, especially the autoimmune diseases.

Simple allergies occur when the immune system launches an attack against what it deems to be a foreign invader, such as pollen, dust, molds, or specific foods. Autoimmune disorders have generally been thought to result when the immune system mistakes some aspect of "self" as foreign and launches an inappropriate attack. But the new model could mean something different. If the system is concerned with danger, then anything that sends a distress signal might attract its attention, and that anything could theoretically be a bioelectrical process or an invading organism. It could be an abnormal alteration in the electrical properties at the cell's surface, for instance—though this is highly speculative at present.

There are more than a thousand autoimmune disorders. One is multiple sclerosis, in which the body attacks its own myelin sheath, an insulating material covering the nerve fibers that is necessary for the normal electrical functioning of the nervous system. (This results in an electrical short circuit so that electrical impulses cannot be carried by nerves.) Another autoimmune disorder is lupus erythematosus, in which the immune system can attack any number of body organs. Still another is rheumatoid arthritis, in which the immune system is directed at the cartilage that cushions bone joints.

No one knows just what causes the body to go after itself, but a problem in both identification as well as the on/off antibody mechanisms is likely. Chemical and viral agents are also suspected triggers, and chronic long-term EMF exposures are possible co-factors given the body of work that has found various frequencies to initially stimulate the immune system, then later suppress it into near collapse. This exact sequence mirrors many immune-system dysfunctions.

Electromagnetic Sensitivity Syndrome

When someone is hit by lightning or receives an electrical shock, electricity courses through the human anatomy along the path of least resistance—the arterial blood supply. It follows the arteries to the heart, then courses up into and moves throughout the brain, which is 70 percent water. Anyone who has received a strong jolt has had bizarre neurological symptoms, such as unusual light phenomena, or seizures, which afterward may never completely disappear.

Although medical practitioners say it is impossible to "feel" low-level electromagnetic fields, many people report otherwise. Some tell of a kind of mild constriction in the upper sinus and forehead region when they are under high-tension lines—a palpable sensation that they find difficult to describe. Others have a desire to get away from such places as radio or TV studios or feel uncomfortable in the electronics or TV departments of stores. Some people are more sensitive than others; some animals are, too.

Others unfortunately report a kind of allergic sensitivity to some frequencies that is being investigated by a handful of researchers in America and in Europe, where electric allergy is clinically recognized. Called electromagnetic sensitivity syndrome, it is occurring in different professions with high EMF exposures: computer operators, airline personnel, and operating-room doctors among them. It may play a role in what is now known as Gulf War Veterans Syndrome.

Once wholly dismissed as psychological in origin, electromagnetic sensitivity syndrome has prompted several medical practitioners to take a clinical interest in the subject for the simple reason that more and more people are reporting the development of allergiclike reactions to different frequencies. There are even indications that people can become sensitized to specific frequencies the way they are to specific allergens like ragweed pollen. But, unlike more easily recognized allergens, those with electromagnetic-frequency reactions also appear to develop cross-reactions to a range of chemicals typically found in perfumes, cleaning fluids, solvents, petroleum products, diesel fumes, sulfur, formaldehyde, and paints, among others.

The reactions can be triggered by the chemical first, with sensitivity to certain frequencies following, or vice versa. It appears to depend on which exposure is the greatest, for the longest period of time. The onset is usually abrupt after the first exposure to a novel field. Symptoms typically include rashes, flulike indications, nausea, dizziness, headache, and sometimes a low-grade fever with swollen glands, as well as a range of neurological symptoms, which can include sound and light sensitivity, concentration problems, vision disturbances, general malaise, and debilitating fatigue.

These can be the symptoms of any number of illnesses, with the exception of one thing. Those who report being sensitive to electromagnetic fields also report becoming increasingly sensitized to a range of EMF devices. A person who has become sensitized to a particular computer, for instance, may react initially by developing a red rash on the neck and face or some other exposed body area after a few hours of use. The next exposure will bring on the reaction sooner, and sooner again with each new exposure. Often these persons then notice that their televisions, stereos, portable phones, or even their regular phones also bother them in a way they never did before if they are within a few feet of them. Some people also develop sensitivities to sunlight afterward, too.

As with all allergies, there is a wide range of severity among different people. Some report symptoms severe enough with repeated exposure to cause debilitating fatigue, depression, decreased memory, sleep disturbances, disorientation, unusual behaviors, and even convulsions. Symptoms often gradually dissipate when the exposures are stopped.

Dermatologists are seeing more and more patients with computer-generated rashes. No one is sure why this is happening, or even what the precise mechanism is in the body's immune system that is being activated, but it will probably prove to be a neurological interaction with the immune system. Unfortunately, treatments are experimental and available at only a handful of clinics. The best advice is to reduce one's exposure to the generating sources. Some doctors recommend moving to a rural environment, where there are fewer EMF exposures, and limiting any contact with known sources within one's control.

Dr. Cyril Smith and Dr. Jean Monro, in England, have set up clinics to test and treat electrically sensitive multiple-allergy patients, as has Dr. William Rea at the Environmental Health Center in Dallas, Texas. Dr. Monro had a number of patients who lived near high-tension lines who became increasingly sensitive to certain frequencies and developed chemical allergies as well.

Dr. Rea himself became hypersensitive to electromagnetic fields during his years as a surgeon. The high-tech operating rooms in which he worked can rightly be classified as a hazardous multifrequency environment. He subsequently discovered that he was not alone in developing a range of neurological and allergic reactions; others were reporting the same problems.

Dr. Rea left surgery and established the Environmental Health Center in Dallas, perhaps the best (if not only) clinic of its kind in America. Patients can be tested in specially shielded rooms to a wide range of frequencies and reactions can be recorded, as well as verified through responses in the patient's autonomic nervous system. Most patients show consistent sensitivities to specific frequencies. Dr. Rea has established electromagnetic sensitivity as a bona fide clinical syndrome in America.

Drs. Smith and Monro, at various locations in England, including Wellington Hospital, Nightingale Hospital, Lister Hospital, and the Hemel Hemstead clinic, have also tested electrically sensitive patients in shielded rooms. They have found that most had electromagnetic sensitivities that were critically dependent on frequencies extending from the millihertz to the gigahertz range. Reactions occurred only if exposures exceeded a certain threshold that was specific for each individual. In other words, those already sensitized had specific frequencies that triggered reactions, but only above a certain field strength.

Having observed that many food, mold, and chemical allergies can be "turned off" with different doses of the same offending allergen (called a neutralizing dilution), Drs. Smith and Monro have successfully treated some electromagnetically sensitive patients by specifically tailoring coherent frequencies to the individual patient. Much of this work is considered outside of mainstream medicine, but Dr. Monro directs her own hospital for the environmentally ill called Breakspear Hospital in Hertfordshire, England, where some people are finding relief from their baffling symptoms.

Many naturopaths are knowledgeable about the use of neutralizing doses of allergens, in place of antihistamines, to treat allergies. This is an interesting method of dealing with an erratic immune system. Details of treatments for electric allergies can be found in *Electromagnetic Man: Health and Hazard in the Electrical Environment,* by Dr. Cyril W. Smith and Simon Best.

As a curious aside, Dr. Smith has also noted that many such patients, when they are reacting allergically to electromagnetic fields, actually emit higher fields themselves. Several have reported various kinds of equipment failure whenever they were close to such devices as robotic systems, electronic ignition systems, and computer-guided cars, all failing each time the allergic person tried to use them.

Chronic Fatigue Immune Dysfunction Syndrome

Chronic Fatigue Immune Dysfunction Syndrome (CFIDS), also known as Chronic Fatigue Syndrome (CFS), is a baffling disorder first reported around 1982 after some one hundred cases became known in the Lake Tahoe area of California. Although no one has an exact count, approximately 2 million people may now fit the Centers for Disease Control strict criteria for clinical classification and be affected by the disease. In England, it is called myalgic encephalomyelitis; in Japan, low natural killer cell syndrome. In the United States, it has been called Epstein-Barr virus and chronic mononucleosis—and also goes by the cavalier epithet of "yuppie flu." Similar syndromes may go back hundreds of years.

No one knows whether CFS is caused by a new agent or an old one manifesting in a new way, but one thing has become abundantly clear to clinicians all over the world: CFS is not a garden-variety absence of energy or pep that is causing previously healthy, happy people to take to their beds for as much as two years. Sufferers from CFS report the sudden onset of mild flulike symptoms, which never quite go away but develop instead into a multisystem disorder of differing severity. Its hallmark symptom is a debilitating fatigue beyond anything the person had heretofore experienced. People with CFS typically say that one day they felt fine and the next day they could hardly pick up their toothbrushes, so profound was the fatigue and muscle weakness. They add that it is not at all like a normal flu, in which the immune system slugs it out with a viral agent for a few days, producing a high fever and the like, after which the sufferer feels drained but better. They describe CFS as more like entering the twilight zone.

Other symptoms, which can come and go for months or years, include low-grade fevers, sore throats, tender or swollen lymph nodes, heart irregularities, an inability to concentrate, mental disorientation, a range of neurological symptoms that can even mimic psychosis, visual and sleep disturbances, abnormal weight changes (either too much or too little), profound muscle pain, and weakness, as well as a return of the debilitating fatigue within twenty-four hours after minimal exercise that in the past was easily tolerated. Depression often accompanies CFS, but in most studies it has been found to result from, rather than precede, the disorder. Environmental illness or a host of allergies and chemical sensitivities often develop within the first year of onset, including in some cases sensitivity to electromagnetic fields.

There are several theories about CFS, the most likely being that it is caused by a retrovirus (a single strand of RNA that must trick the nucleus of a cell into incorporating it into its normal DNA manufacture), which infects the immune system's B cells not unlike the way the AIDS retrovirus infects the T cells. Some think the virus lives mainly in brain tissue, thereby eluding most tests. Others think it is a deep viral infection of the liver, which along with the thymus gland is the seat of the body's immune system.

Although the agent or agents have not yet been identified, researchers know several things about the syndrome. Something causes the immune system of CFS patients to chronically underreact and then severely overreact. Often it seems as if the immune systems of those with CFS is on constant alert. But no one knows whether this occurs before or after CFS sets in. Some viruses in the herpes family have been found at abnormally high elevations in those with CFS, including the Epstein-Barr virus (a majority of the population has been exposed to this virus and will test positive for it by the adult years). But it is not known whether these abnormal elevations are primary infections or are a manifestation of normally quiescent infections kept in check by a previously healthy immune system that goes haywire when an immune dysfunction sets in. Most researchers suspect the latter, since the Epstein-Barr virus has been around for a long time. It is also widely felt among clinicians who have treated CFS patients that the allergies they develop are not typical ones, but are based in some neurological or immune-system problem.

Unfortunately, certain cancers associated with the Epstein-Barr virus are beginning to turn up in the CFS population. Cancers of the pharynx and the back of the nose (called nasopharyngeal carcinoma), cancer of the salivary glands, thymic carcinoma (a deadly cancer of the thymus gland), and primary intracerebral lymphoma (an immune-system cancer of the brain) are being found in higher than expected numbers. The Epstein-Barr virus has a predilection for the immune system's B cells, and B-cell carcinomas, a kind of lymphoma, are also turning up. B-cell carcinomas differ from other lymphomas such as Hodgkin's in that they have a strong tendency to infect the central nervous system.

Are there connections between CFS and electromagnetic fields? It is likely that EMFs are a co-factor in the disease process, if not the outright initiator. CFS is widespread in the electronics industry in California's Silicon Valley and along Boston's Route 128 corridor. It affects women far more often than men. (Among the highest number of sufferers are computer operators, who typically are women sitting for hours in fields that directly impact on the thymus gland, located near the thyroid in the lower neck behind the sternum.) More than a hundred Silicon Valley employees have filed lawsuits against their companies for the condition alone. Many other companies grant sick leave for employees diagnosed with CFS, and the Veterans Administration grants disability pay for afflicted veterans.

Whether CFS will in time be definitively linked with electromagnetic fields remains to be seen, but it bears striking similarities to electromagnetic sensitivity syndrome. And many observations regarding EMFs and the human anatomy appear to have the same imprints as CFS. Given the research on electromagnetic fields and immune-system activation followed by long-term immune suppression, an extrapolation to CFS is not out of line. Also, consider the findings on EMFs and nighttime melatonin suppression, and relate them to the common paradox in CFS patients that, despite their overwhelming tiredness, they are characteristically unable to sleep between 2:00 and 5:00 A.M. A link is suggested. EMFs are associated with a range of hormonal/neurological effects and central nervous system problems. Plus the research indicating EMFs as co-factors in human disease states as well as viral mutations cannot be ignored.

Gulf War Syndrome,
the Military, and Electromagnetic Pulse

A back door into understanding both CFS and electromagnetic sensitivity syndrome may have been unwittingly provided by the U.S. military in the Persian Gulf War. It wouldn't be the first time that the military has provided an unintentional epidemiological survey of uninformed test subjects. Several years ago, the Army tried to investigate the effects of lead poisoning on the reproductive health of artillery personnel who had handled metal casings. The findings from the control group used in comparison surprised the investigators. Their sperm counts were even lower than those exposed to lead! The control group had in fact been regularly exposed to radio-frequency microwaves, which clearly had an affect on their reproductive abilities.

Veterans of the Persian Gulf War are reporting illnesses that closely resemble CFS. The disorders may have been triggered by several things, including chemical pollution from the burning oil fields, a parasitic infection transmitted by sand fleas called leishmaniasis, and possibly electromagnetic fields used in modern weaponry as well as in some experimental weapons reportedly used for the first time. The forces were also exposed to a range of lethal chemicals released when Iraqi chemical factories were bombed and possibly were exposed to biological warfare agents, which the Iraqis were rumored to have used. In addition, there is reason to believe that mustard gas may have been used, as some of the European forces in the Gulf found traces of it afterward.

At least 20,000 American servicemen and veterans are suffering from a constellation of symptoms well known to those with CFS. The symptoms again point to a multisystem disorder affecting the immune system. Complaints include debilitating chronic fatigue, skin rashes and sores, an inability to concentrate, memory loss, headaches, heart irregularities, severe muscle pain, gastrointestinal difficulties, shortness of breath, and chemical sensitivities that they never had before. In addition, the wives of some Gulf War veterans are experiencing similar symptoms and a range of reproductive difficulties, including increased miscarriages. Some children conceived after the spouse's return are said to be more sickly and to have a host of chronic health problems, in addition to an increased number of birth defects, especially of the urinary tract. The implications are that the causal agent, if biological, is contagious to the spouses and is mutagenic to their children.

Many of the symptoms are similar to those of Vietnam veterans who were exposed to Agent Orange, especially the skin rashes and sores. It is possible that a combination of many exposures in the Gulf War is causing immune-system problems. And it may very well turn out that today's high-tech electronics—common in both those wars for the first time in history—also play a role.

Many military positions in all branches of the service today require that soldiers operate in high-intensity multifrequency fields a majority of the time. We have become completely reliant on sophisticated gadgetry. It directs everything from communications apparatus to missile guidance to devices for jamming enemy systems. But electronic equipment can be easily disrupted by other electromagnetic devices. There is some indication that as many as five state-of-the-art Black Hawk helicopters and perhaps twenty-two crew members may have been lost due to electromagnetic interference when the helicopters flew too close to high-tension lines or military radio transmitters. (Such accidents have decreased since 1987, when the military initiated a program to shield the helicopters and to warn pilots about EMF sites.)

Electromagnetic interference may be the hidden problem behind many of the "friendly fire" American casualties—deaths of Americans caused by other Americans. In the 1986 U.S. attack on Libya, out of thirty-three high-tech American fighter planes, one plane crashed mysteriously and seven were unable to fire a single shot. The reason given was the electronic blizzard from America's own high-powered military transmitters, designed to jam Libya's antiaircraft defenses, seek out targets, guide weapons, and communicate with each other. It is a likely suspect in many other such accidents and a problem that the military is well aware of.

The Pentagon has also known for several decades about a phenomenon called electromagnetic pulse, or EMP. EMP is a massive overdose of energy to a given area at any one time, such as a by-product of a nuclear explosion. But it can also be created by simple electrical means and used in a weapon. It can and will interfere with any electrical signals in operation. Just as in the 1951 science-fiction movie classic, *The Day the Earth Stood Still*, starring Michael Rennie and Patricia Neal, everything in the path of EMP comes to a screeching halt. All telephone, radio, satellite transmission, electrical transmission—therefore all modern appliances and devices—and anything motorized, like elevators, cars, or such military equipment as tanks and planes, simply cease to function. This could include airplanes in flight and hospital equipment. EMP will also scramble computer information stored on floppy or hard disks, whether the computer is plugged in or not. (If it can do that to electronic hardware, imagine what it can do to human software.)

A single nuclear explosion just one hundred miles above Kansas City would create an EMP that could knock out all communications and electronic devices across the whole United States. How long it would last is speculative, but almost any electrical equipment that was plugged into a socket at the time would be completely destroyed. The electric lines would be, too. An absence of communications could, therefore, last as long as it takes to replace equipment and all electric lines, including electric transmission and distribution networks—in other words, months if not years.

This situation is obviously serious in a world where small nuclear devices may actually be available to unstable countries with ill intent toward their global neighbors. It turns out that a country doesn't need a whole arsenal to incapacitate another nation; just one nuclear weapon detonated high in the atmosphere will do.

In an attempt to offset an EMP-war scenario, the military created the Ground-Wave Emergency Network (GWEN), a communications system to operate in the very low frequency range between 150 and 175 kilohertz with radio waves that hug the ground. GWEN stations transmit in a 360-degree circle to the next transmitters. The original system was supposed to have 86 transmitters, each 300 to 500 feet high, stationed every 250 to 300 miles across the country. Approximately 50 were built before military cutbacks defunded the remaining towers. Changes in political administrations could continue the program in the future.

GWEN transmitters, each with a peak power output of 2,000 to 3,200 watts, broadcast in "standby" mode approximately one percent of the time and subject the civilian population to limited, though regular exposure. Some GWEN sites are suspected in electromagnetic interference with other electrical equipment. In 1993, a branch of the National Academy of Sciences, the National Research Council, issued a report that concluded there would be a "minimal" risk of "less than one" additional death over a 70-year period for persons living within ten kilometers of the GWEN system. But the NRC risk-assessment model came under scrutiny as being too low. Critics noted that in order for cancer trends to be obvious in epidemiological studies, increases of fivefold or greater would need to be routinely observed. With so few studies available of people living near radio and TV transmitters, the model that the NRC committee used to reach their minimal projection may be inaccurate.

The NRC committee was surprised, by all reports, by the absence of appropriate data upon which to base their conclusions and noted that such studies were needed. Their report also carefully distinguished GWEN systems from other commercial broadcast facilities (radio and TV), lest the population begin to question their safety too. Critics say that the whole notion of the GWEN system is specious and is based on the false notion that nuclear war is winnable, which it isn't. There is evidence that the system itself would not survive an EMP and therefore has no reason for existing.

The U.S. military has been experimenting with doomsday scenarios for a number of years, conducting EMP tests to simulate the effects of nuclear blasts on communications by generating 50,000-volt "zaps" in various research programs. One such program, called Empress II, was operated under the U.S. Navy from 1988 until recently. Empress II was designed to simulate EMP on ships and tested at different times in the Atlantic Ocean off the coast of North Carolina. Empress II was recently scrapped due to military cutbacks, but other military EMP programs remain intact.

Unfortunately, some military contractors may have been experimenting with EMP on their employees, several of whom have died from different forms of leukemia. Some defense workers have won workers' compensation claims against their employers. One such suit (*Strom v. Boeing*), filed by 27-year Boeing employee Robert Carl Strom, resulted in a $500,000 out-of-court settlement in 1990, in which Boeing also agreed to set up an ongoing medical program to monitor the health of similar workers there. Strom, whose job involved testing EMP on MX missile parts, firing high-voltage EMP simulators hundreds of times a day inside an enclosed room, claimed that he contracted chronic myeloid leukemia in his top-secret work there, as did several of the 700 other workers in the Boeing program. During the suit, Strom discovered that Boeing had given data to the Lovelace Biomedical and Environmental Institute of Albuquerque, New Mexico, which enabled Lovelace to study the effects of EMP on humans.

It has been rumored for several years that the military has developed a pulse gun, and that the Navy used nonnuclear electromagnetic pulse weapons during the Gulf War to disrupt and destroy Iraqi electronic communications and defense systems. EMP warheads were said to be mounted on some Navy Tomahawk cruise missiles in the early days of the war. The U.S. Army also has a high-powered microwave (HPM) device on tanks, which clears mines by destroying their circuitry.

Although the military has denied using such EMP weapons against Iraq, some military sources report that Tomahawk-size

EMP warheads are being developed at Los Alamos National Laboratory in New Mexico and at Eglin Air Force Base in Florida. And others report that EMP weapons are ultimately intended for use against humans, in that they have the advantage of being completely silent, imperceptible at first, and nonlethal if so intended. These "advantages" make them a tempting option for civilian crowd control, for terrorist groups, and for the prevention of security breaches at military installations.

A report from the Microwave Research Department at the Walter Reed Army Institute of Research said that the range of 1 to 5 gigahertz was a militarily important range because it penetrates all organ systems of the human anatomy and therefore puts all organ systems in jeopardy. Effects on the central nervous system are of particular interest to the military. The report was based on a testing program begun in 1986 that was divided into four areas of study: prompt debilitating effects, immediate stimulation through auditory effects, work stoppage effects, and stimulated behavioral effects. The report found that microwave pulses appeared to couple with the central nervous system and produce effects similar to electrical stimulation unrelated to thermal reactions. Much of this military research verifies José Delgado's work on altered human behavior and EMFs.

It may take years to unravel what veterans of the Gulf War were actually exposed to in the form of chemicals, biological agents, and electromagnetic fields. But given the research and case histories that already exist for electromagnetic sensitivity syndrome contained within clinical environmental medical literature, as well as that for chronic fatigue syndrome, it may be possible to piece some of it together before all the component parts are understood.

It may turn out that all three syndromes—chronic fatigue, electromagnetic sensitivity, and Gulf War—are the same immune-system dysfunction, caused by similar interactions between EMFs and chemical stimulants but manifesting itself in slightly different forms.

AIDS

Acquired Immune Deficiency Syndrome (AIDS) is caused by the human immunodeficiency virus (HIV). HIV is a retrovirus, a single strand of RNA that infects the immune T cells and tricks the DNA into manufacturing new virus. It is almost as if the virus is invisible to the immune system, so much of the cell's regular apparatus does it enlist for its own purposes. HIV is truly a Machiavellian life-form. It mutates quickly, making the probability of a vaccine unrealistic, and some estimate that there are now a few hundred different strains in various parts of the globe. It may eventually kill over a third of the world's population.

HIV seemed to come out of nowhere. No one knows whether it is a new virus or an old one that has mutated into a newer and deadlier form. It is certainly curious that viruses with a particular penchant for the immune system, like HIV and the as-yet-unidentified virus that probably causes CFS, appeared at the same time.

One aspect of the HIV virus that has not been widely discussed is that it shares some characteristics with viruses known to cause various forms of leukemia in both humans and animals. At the Harvard School of Public Health, Dr. Julie Overbaugh and a research group found a cat-leukemia virus capable of producing either feline leukemia or a full-blown acute immunodeficiency syndrome that acts like human AIDS. The researchers were able to create a mutation in the virus, which then caused the fatal immunodeficiency syndrome and quickly killed the experimental animals. Their conclusion was an interesting one — that a subtle mutation could alter a minimally pathogenic virus into a deadly one capable of infecting and collapsing the immune system.

Are we perhaps witnessing the same dynamic with the HIV virus? Was it once relatively nonlethal? Has it mutated into an incurable form? EMFs are known to cause genetic mutations. Is there a link? At the very least, EMFs adversely affect the immune system. Are those who develop AIDS already in a state of immune suppression when they become infected with HIV? Is pre-existing suppression a necessary condition for infection? It would appear so, because HIV is not easy to get. There is ample evidence that, even with an unsafe lifestyle, repeated exposures are often necessary for transmission of the virus. And once HIV has taken hold, chronic exposure to EMFs may play a significant role in the suppression of the immune system, thereby tilting the balance from a person being HIV-positive to developing full-blown AIDS.

Some research exists regarding EMFs and T-cell lymphocytes. Dr. Daniel B. Lyle, at the Jerry L. Pettis Memorial Veterans Hospital in Loma Linda, California, found that human T-cell lymphocytes in culture, when exposed to low-strength 60-hertz electric fields for only forty-eight hours, showed a significant reduction in their ability to combat foreign cells. Although this represents only a small body of research, the findings are indicative of important negative interactions between normal power-frequency fields and the immune system's T cells. Anyone who is HIV positive may want to pay close attention to all possible EMF-mitigation strategies. It can't hurt, and it may help.

Amyotrophic Lateral Sclerosis

Amyotrophic lateral sclerosis (ALS), also known as "Lou Gehrig's disease," after its most famous American victim, is a progressive degeneration of the nerve cells that control voluntary motor function. Victims of this motor-neuron disease (an electrical disorder) eventually lose all ability to move or speak, but, cruelly, the mind remains untouched, along with general sensation, bladder control, and eye motion. The brilliant English physicist Stephen Hawking has lived with the disease for 20 years. Other notables who have been ALS victims include actor David Niven, U.S. Senator Jacob Javits, and composer Dmitry Shostakovich.

ALS is considered a rare disease, with a worldwide incidence of between 0.4 and 1.8 cases per 100,000 people. Approximately 5,000 cases are diagnosed annually in the United States, and it is more common in white males in their fifties and sixties, only 5 to 10 percent of whom have a family history of the disease.

A hereditary form of ALS typically strikes younger people. Of those with the hereditary form, about half have been found to have mutations in chromosome 21, the gene that encodes for superoxide dismutase (SOD), an enzyme that helps protect the body against damage from unstable oxygen molecules commonly called free radicals. It is unclear if the same defective SOD gene plays any part in the 90 percent of non-inherited ALS cases.

Other researchers have noted that ALS victims have too much glutamate, an amino acid that serves as a chemical messenger between nerve cells. Too much can kill motor neurons. And still other research points to ALS being an autoimmune disorder.

While the cause of ALS is unknown (some agricultural chemicals and pesticides are suspected), an unpublished epidemiological study done in 1986 by Dr. Zoreh Davanipour, at the University of Southern California, found a link between EMF occupational exposures and ALS. Airline pilots, welders, electricians, and other electrical workers were found to have increased incidences of ALS after twenty-year exposures classified as high (above 10 milligauss, or above 100 milligauss for intermittent exposures), medium (above 5 milligauss), and low (less than 1 milligauss).

Also, an ALS cluster was identified in neighborhoods near a powerful Federal Aeronautics Administration (FAA) air-traffic-control radar installation near South Patrick Shores, Florida. Eight people in a community of only about 2,000 have been identified as having ALS since 1980—nearly twenty times what would normally be expected. The same residential neighborhoods also have been found by Florida's Department of Health to have a threefold increase in Hodgkin's disease and elevated rates of breast and cervical cancer.

Based on such limited data, speculation about a causal relationship between radar or other frequencies and ALS is premature. But any occupational tendency or disease cluster involving this rare disease and EMF sources ought to pique the curiosity of researchers.

Alzheimer's Disease

Many people think that Alzheimer's disease—a progressive, fatal, complete mental deterioration—is a disorder of normal aging that afflicts an unlucky few. But there is nothing about Alzheimer's that is normal to aging. Nor is it related to mild forgetfulness. It is a degeneration of neurons in specific areas in the brain that results from some disturbance within nerve-cell networks utilizing the neurotransmitter acetylcholine.

Just ten years ago, Alzheimer's was considered an obscure and rare condition, but today it is the nation's fourth leading cause of death. What happened? Is it simply that better diagnosis has turned up more statistically reliable numbers, which perhaps had been lumped together in years past with senile dementia? Or are we dealing with another degenerative nerve disease increasing in incidence beyond a mere increase in the population? It looks like the latter is true—although increases in the population of those living beyond the age of eighty-five plays a significant role in the sheer numbers of cases today. And although there are only a handful of indicative studies and much speculation at this stage, there is a possibility that some EMF frequencies may play an important role, too.

Alzheimer's is a specific organic disease that afflicts only some people. It is quite different from the memory lapses that plague all of us as we age, in which long-term memory is crystal clear and short-term memory seems to all but evaporate. A typical memory lapse of old age would be a person's remembering in vivid detail an event from youth as if it were yesterday but forgetting where his or her glasses were a minute ago. With Alzheimer's, people forget they ever wore glasses.

Alzheimer's is a physical process in which the nerve cells of the brain take on the abnormal characteristics of "plaques" and "tangles." In time brain tissue comes to resemble long strands of gray knotted rubber. The disease affects women twice as often as men. Women who have taken anti-inflammatory drugs for arthritis or have had estrogen-replacement therapy have been found in some studies to have a reduced risk of developing the disorder. These studies indicate that inflammation as well as hormonal changes may be important factors. (EMFs and hormonal changes were discussed in Chapters 7 and 8.)

Research particular to acetylcholine was conducted in 1976 by a research group headed by J. J. Noval, at the Naval Air Development Center in Johnsville, Pennsylvania. Studies using rats exposed to very weak electric fields vibrated in the extremely low frequency ranges (the kind of EMFs typical of any office or modern home) produced an increase in brain-stem acetylcholine levels, indicating a subliminal stress response in test animals. (This also has implications for humans and low-level "contact currents" produced by touching any common machine, including small appliances. Far more work needs to explore this possibility.)

Genetics may also be involved. Several studies have found a genetic abnormality similar to those with Down's syndrome also occurring in Alzheimer's patients. And recent research has found that the presence of a protein molecule called apolipoprotein E (ApoE4) was present in 64 percent of those studied with Alzheimer's, whereas only 31 percent of those in the control group had E4. (However, the presence or absence of E4 was found to have a clearer relationship to the age of the person at the onset of Alzheimer's. Some of those without E4 did get the disease, but the onset was after age eighty-four.)

Recent research by Daniel Alkon, at the National Institute of Neurological Disorders and Strokes, has turned up a fundamental difference between the skin cells of Alzheimer's patients and healthy people. Alzheimer's patients appear to have defective potassium ion channels, which funnel potassium out of the cells. It was found that Alzheimer's patients had this cellular malfunction in the nerve cells leading from the nose to the brain. Learning and memory are associated with a number of changes in the flow of potassium ions through cellular channels. It is not known yet whether the defect originates within the brain, or even whether it precedes Alzheimer's symptoms. All that remains to be seen.

Some important questions need to be asked, such as: Are different EMF frequencies responsible for opening and closing (or permanently shutting down) potassium ion channels in the same way that research indicates window effects for calcium ion channels at the cellular level? Could an EMF resonance factor be involved with potassium ions? Melatonin is also known to be suppressed in those with Alzheimer's, and EMFs have been found to lower melatonin in some studies. Is there any significance to the concentration of magnetite in the nasal area? What of the studies that have found EMFs to increase the permeability of the blood-brain barrier?

Recent work done jointly by Dr. Eugene Sobel, of the University of Southern California School of Medicine, and Dr. Joseph Bowman, of the National Institute for Occupational Safety and Health, found statistically significant increases in Alzheimer's in some EMF-related occupations. The researchers combined data from one American study and two Finnish studies and found that tailors, seamstresses, and dressmakers (who work with electric sewing machines) were overrepresented among the Alzheimer's cases. Increases were also seen for carpenters and electrical engineers, among other EMF professions. A fourth study is in the offing, as well as additional research in Finland.

Dr. Sobel indicated that the use of certain high-EMF-emitting machines may eventually be linked with Alzheimer's, but that a causal relationship between EMFs and specific people is premature. Kitchens, however, are high-EMF sources, and this may eventually account for the two-to-one ratio between women and men with Alzheimer's.

There is also some indication that the microwave frequencies are particularly suspect. Dr. Sam Koslov, director of the Applied Physics Laboratory at Johns Hopkins University, found, in a study using microwave exposures on chimpanzees, that repeated low-level nonthermal exposures to the eyes produced clinical Alzheimer's in test animals. At autopsy, the classic plaques and tangles were found in brain tissue. (The researchers discovered this relationship by accident; they were testing for something else.)

Regarding the causes of Alzheimer's, a range of possibilities exists, including subtle genetic alterations initiated by environmental EMFs. Or EMFs may be acting as co-factors in melatonin suppression and in changes to the blood-brain barrier, potassium ion channels, or acetylcholine levels in the brain stem, among other possibilities. With between 2 and 4 million people afflicted with Alzheimer's in the United States alone, this will prove to be one of the most provocative research areas within the next few decades.

Cancer

As already mentioned, the overall incidence of cancer has been rising steadily since the turn of the century despite earlier diagnosis, better surgical techniques, and improvements in treatments. Viral infections are thought to account for up to 20 percent of some cancers; environmental factors contribute perhaps the remaining 80 percent.

It isn't possible to devote entire subsections in this work to all the various forms of cancer, but it is possible to discuss certain types of cancer that appear to be more related to EMF exposure than others. These would broadly include cancers of the brain, the blood, the skin, and glandular tissue.

As discussed in Chapter 8, cancer is the wild proliferation of cells that were at one time in a normal state. It is thought that cancerous cells are stuck in a partial state of dedifferentiation, that something has gone askew with the growth-triggering mechanisms within the cell itself. EMFs in various frequencies have been found to cause the kind of genetic damage that could lead to such abnormalities; to influence cell division at a crucial stage of mitosis; and to promote an increased rate of growth in cells that are already tumorogenic. Other frequencies at higher power levels have been found to shut off tumor growth. This remains an exciting research area for the identification of both physiological hazards and potential medical therapies.

The following sections describe some of the studies (there are hundreds) that have found associations between EMFs and cancer. There are gigantic holes in the research regarding certain frequencies, not all studies have been replicated or their conclusions verified, and it will probably take scientists many years to get a complete biological understanding, but here is a glimpse of the picture that is emerging.

Brain Cancer

A growing body of evidence indicates that brain tumors may actually be markers for EMF exposure. About twenty studies have found associations between EMFs and brain tumors, with ten studies showing strong statistically significant increases in brain tumors in high-EMF occupational exposures or in the children of such workers. Another eight studies are currently under way as of this writing.

The incidence of primary (first site) brain tumors has risen threefold since the 1960s. Mortality from brain cancer in white men and women nearly tripled between 1968 and 1983. Primary brain tumors have always been considered extremely rare. Until recently, most cancer in brain tissue had metastasized via the bloodstream from cancers in other areas of the body, such as the breast or the liver. The distinction is an important one, because it may help lead us backward toward the dynamics causing cancer in general.

Whether primary or metastatic, brain cancers can occur anywhere in the brain, in people of any age. The causes are completely unknown, with genetics thought to play a minimal role at best. Symptoms can have great diversity and are even independent of tumor type from individual to individual.

The most common form of brain cancer is a glioma, a tumor that occurs in the supportive tissue of the brain. Astrocytomas are the most common subtype of gliomas, and it is astrocytomas that have been found to be significantly elevated in people exposed to higher levels of EMFs for more than five years in some studies. Telephone-cable splicers and certain electrical workers were found to have as high as a 70 percent increased risk of brain tumors. Cancer of the meninges, the layer that surrounds the brain and has a high concentration of magnetite, has increased as well.

Children may be particularly susceptible. Those living near high-current power lines showed a doubling of brain-tumor risk in some studies (although not in others), and children whose fathers were employed in EMF-related occupations had a 60 percent increased risk of central nervous system/brain cancers. The children of electricians had an increase of three and a half times over the expected rate, with a predominance of tumors in the brain stem. These studies, along with others, seem to indicate that EMF damage in parents occurs in the germ or sex cells, meaning it is transmissible to the next generation.

Dr. Samuel Milham, Jr., has done extensive research on EMF exposure and occupational risks. In one study, he found that electricians had a 55 percent increased risk of developing brain tumors, and for all high-exposure occupations the increased risk was 23 percent.

Many of the occupational studies have found the length of exposure and the dosage to be factors in tumor development. In one study, the risk for astrocytomas among those in electrical professions rose according to the duration of employment. At five years, the relative risk was 3.3; at five to nineteen years, it rose to 7.6; at twenty years and over, the relative risk of developing an astrocytic tumor was 10.4.

A 1993 utility-sponsored study in Canada found that electrical workers with the longest cumulative exposures had twelve times the expected rate of astrocytomas. And an Australian study found that women who worked with computer monitors (cathode-ray tubes) had five times the expected rate of primary gliomas. (This was the first study to link computer use with brain tumor incidence.)

Until recently, dosage and duration of exposure were thought to be important factors only with ionizing radiation. Cumulative effects from the nonionizing band, if any, were considered transient and therefore unimportant. That there are cumulative effects for EMFs could be among the most important findings to date, with serious implications for our "innocuous" everyday exposures.

Leukemia: Cancer of the Blood

Leukemia is cancer of the blood, and there are several forms, some more deadly than others. While leukemia is thought to be more common in childhood, in fact a far larger number of adults contract it. In America, approximately 24,000 new cases of leukemia are reported each year, about 21,500 in adults and 4,000 to 5,000 in children, with deaths numbering around 16,000 annually. Radiation, chemicals, genetics, and some viruses are suspected causes.

Leukemia is characterized by the production of abnormal, immature white blood cells called leukocytes, which, as the disease progresses, interfere with the healthy white blood cells (granulocytes) that the body needs to fight bacterial, viral, and other infections. Red blood cells, responsible for oxygen transport throughout the body and blood platelets needed for clotting, are also affected. As healthy white cells decrease, infections set in; as healthy red cells decrease, anemia and bleeding disorders take over. In time, all organs of the body are affected.

The two basic types of leukemia are acute leukemia, marked by a great number of immature blast cells, and chronic leukemia, marked by elevations of both immature and mature white cells. As white cells proliferate beyond what the body needs, normal bone-marrow elements are crowded out and symptoms become systemic. Acute and chronic leukemias are further classified depending on whether they originate from granulocytes (in which case they are called myeloid or myelogenous) or from lymphocytes (in which case they are called lymphocytic). The two most common forms of acute leukemia are acute myeloid or myelocytic leukemia (AML), which occurs most often in adults; and acute lymphocytic leukemia (ALL), which occurs mostly in children. The incidence of the acute and chronic forms is about equally divided.

Acute myelogenic leukemia, or AML, is especially frightening in its sudden onset and rapid progression, sometimes within only a few weeks. Typical symptoms include fever, fatigue, bone pain, swelling, severe anemia, bruising, bleeding, and infections. If treatment isn't prompt, AML can be fatal within weeks. Acute lymphocytic leukemia (ALL), which primarily affects children under age five, also has those symptoms, but the survival rate is often near 50 percent. Prognosis with adult AML often isn't good. Treatment for the acute forms is intensive and requires highly skilled hospital care. New combinations of drugs, radiation, and bone-marrow transplants are increasing the survival time for some acute leukemia patients.

The chronic forms of leukemia, though less frightening, are serious illnesses nevertheless. Chronic myelogenous leukemia (CML) can come on gradually and most often occurs in men between twenty and fifty years of age. Symptoms can vary greatly, ranging from none at all (discovery is often through a routine blood test) to anemia, general malaise, weight loss, night sweats, fatigue, or an enlarged spleen causing discomfort on the left side of the abdomen. A chromosomal abnormality called Philadelphia chromosome (ph1) is present in 90 percent of CML patients. Although CML cannot be cured, remissions are possible, sometimes lasting for years. But CML can also progress to the acute stage; half of those afflicted die within four years of the initial diagnosis.

Another form of chronic leukemia is CLL, or chronic lymphocytic leukemia. CLL most often affects people aged forty to seventy and results from a proliferation of defective white blood cells. Symptoms are similar to those of CML and include decreased antibody production, leaving people vulnerable to infections. The prognosis for CLL is better, however; many people with the disease survive for five to ten years (and sometimes up to twenty years) after the initial diagnosis.

Approximately thirty studies have found either an excess risk or a statistically significant increase in leukemia in relation to EMF exposure, especially in the 50-to-60-hertz range—the one most studied. The increase range for excess risk is considered between 1 to 2 percent above average, while anything over 3 percent is considered statistically significant.

Two Swedish studies published in 1992 sent mild shock waves through the electric-utility industry worldwide. In a study of homes near high-tension lines, Dr. Anders Ahlbom and Maria Feychting, of the Institute of Environmental Medicine at the Karolinska Institute in Stockholm, found that children exposed to average fields of 3 milligauss or more had nearly a four times higher than expected rate of leukemia. An occupational study, headed by Dr. Birgitta Floderus and her colleagues at the Department of Neuromedicine at the National Institute of Occupational Health in Solna, found that men exposed to the same magnetic field range (around 3 milligauss) at work suffered three times the expected rate of chronic lymphocytic leukemia (CLL). Based on these and other studies, the Swedish government became the first in the world to recognize a link between EMFs and cancer risk.

The Swedish studies found the leukemia increases to be incremental and dose related in children. Children exposed to 1 milligauss had twice the expected rate; children exposed to 2 milligauss had nearly three times what was expected; and those exposed to 3 milligauss had close to a fourfold increase. For adults exposed to more than 2 milligauss (as against 1 milligauss), Ahlbom and Feychting found a 70 percent excess risk for acute myeloid leukemia (AML) as well as chronic myeloid leukemia (CML), although this was not considered statistically significant.

Other Scandinavian epidemiological studies have linked EMF exposure and cancer. In male electrical workers, Drs. Tore Tynes, Aage Andersen, and Froydis Langmark, of the Cancer Registry of Norway in Oslo, found a statistically significant increase of 40 percent for acute and chronic forms of leukemia, with the highest risk being among radio and TV repairmen, radio and telegraph operators, and power linesmen.

In addition, French researchers found a fivefold increase in AML among high-EMF occupations. And the same 1993 Canadian study that found the twelvefold increase in astrocytomas among some electrical workers also found that workers exposed to magnetic fields of 2 milligauss or more for ten years or longer had more than twice the expected risk of developing AML. Those with the greatest cumulative magnetic exposures at the top end of the range were found to have up to thirty-eight times the chance of developing AML.

An American researcher, Dr. Genevieve Mantanoski, at the Johns Hopkins University School of Hygiene and Public Health in Baltimore, has done several studies on EMF exposure and occupational health risks. Some of her findings include higher rates of leukemia for power linesmen. And the work of Dr. Nancy Wertheimer and Ed Leeper (previously mentioned in Chapter 9) also found higher risks for children living in high-current households. The Wertheimer-Leeper studies found that the 60-hertz magnetic component at strengths of 3 milligauss resulted in a two- to threefold increase in childhood leukemia and brain cancers. Their work has been verified by several researchers since then.

For other frequencies, such as radio waves, a series of reports on ham-radio operators found statistically significant elevations of AML and other problems of the lymphatic system. Two childhood leukemia clusters have also been identified near military communications transmitters, one in Hawaii and the other in Scotland.

Lymphoma

Lymphomas are cancers of the lymphatic system, particularly of the lymph nodes. Lymph nodes are composed of the immune system's T-cell and B-cell lymphocytes, a class of white blood cell that can become malignant, proliferate, and take over the entire node. Malignant lymphocytes can then spread to other organs of the body.

There are T-cell lymphomas, B-cell lymphomas, and lymphomas that are mixed or composed of both. B-cell lymphomas are the easier to treat, while the more rare T-cell lymphomas, which often involve the skin, are more difficult. Another form of the disease is Burkitt's lymphoma, which is differentiated by its location in the abdominal area, where large masses result in kidney damage and central nervous system involvement. A contagious viral infection is suspected in Burkitt's lymphoma.

Hodgkin's disease is a special form of lymphoma characterized by the presence in the lymph nodes of a unique cell called the Reed-Sternberg cell. The disease spreads predictably from its usual beginning in the lymph nodes of the neck, chest, and armpits to the spleen, liver, and nodes surrounding the aorta near the heart. It is more common in men than women, usually making a first appearance between the ages of fifteen and thirty-five, but sometimes after age fifty as well. If caught early, Hodgkin's often responds dramatically to treatment and has a 90 to 95 percent cure rate. If it has spread, however, treatment is less successful.

Non-Hodgkin's lymphomas are more deadly—and unfortunately more common—than Hodgkin's. Although still considered rare, with between 7,000 and 8,000 new cases reported annually in the United States, by the time the disease is detected it usually has spread. Jacqueline Kennedy Onassis died of Non-Hodgkin's lymphoma in 1994 at the relatively young age of sixty-four despite her otherwise excellent health. The disease had spread to her brain and liver.

Non-Hodgkin's lymphomas can show up in the same way as Hodgkin's, with lymph-gland swelling in the neck, armpits, chest, or groin region, but the symptoms can also include anemia, enlargement of the spleen, general malaise, weight loss, night sweats, and fever. Chemotherapy and some steroids can produce remission in about 50 percent of cases, but the remission is shorter than with Hodgkin's.

Several studies—usually of childhood leukemias—have found associations between lymphomas and EMFs. This is not surprising, since both forms of cancer originate in the body's blood system. In 1992, Dr. Jorgen Olsen found a significant fivefold increase of Hodgkin's lymphoma in children living near 60-hertz high-voltage facilities with exposures of only 1 milligauss or more. Dr. David O. Carpenter, of the New York State Department of Health in Albany, has also found strong associations between 60-hertz electrical distribution lines and increases in childhood cancers (leukemia, lymphoma, and soft-tissue tumors) with exposures of 2 milligauss.

There may be a link between some cancers and ornithine decarboxylase (ODC), an enzyme that is essential for the growth of cells through DNA synthesis. Certain cancer-promoting agents increase ODC activity. Work done by Drs. Ross Adey and Craig Byrus with researchers Susan Pieper and Karen Kartune found that ODC activity increased up to 50 percent in human melanoma cells when lab samples were exposed to a modulated 450-to-500-megahertz microwave field at 16 hertz of only 1 mW/cm^2. The effect was greatly heightened in the presence of phorbol esters, a plant derivative of croton oil. Additional work by the same researchers found that a one-hour exposure to 60-hertz fields produced a fivefold increase in ODC activity in cultured lymphoma cells. ODC activity was also increased up to threefold in mouse myeloma cells (a type of plasma cell) after an exposure of only one to two hours. It was short, intermittent exposures to 60-hertz fields that produced the elevations.

There are numerous lawsuits in various stages of deposition linking power-line and radar frequencies with different cancers, many of which affect the blood, lymph, and soft tissue. Several hand-held police radar gun cases involve non-Hodgkin's lymphoma, melanomas, acute leukemias, thyroid and testicular cancers that officers say are occupationally caused by proximity to radar units within their cars. A few suits also have been brought on behalf of people with non-Hodgkin's lymphoma who live near high-tension lines. It is likely that a link with EMFs, either as initiators or as co-promoters, will eventually be found, perhaps through ODC activity.

Melanoma

Malignant melanoma, the deadliest form of skin cancer, is the fastest-rising cancer anywhere today, with dramatic increases since the 1950s. Prior to 1955, melanoma was quite rare, but from 1975 to 1992, cases in the United States alone tripled. Today, 32,000 Americans are diagnosed with melanoma annually, and 6,800 die. The figures are comparable in other countries with fair-skinned populations. Not only is melanoma considered pandemic; it is also affecting younger and younger people.

For years, dermatologists thought melanoma was related almost exclusively to exposure to the sun's ultraviolet-B (UVB) rays. Sunscreen manufacturers produced products to block that part of the spectrum, and just about everyone thought they were protected when in the sun. Recent research is currently focusing on the UVA component of sunlight as well, but many questions remain when melanoma is only approached from this point of view. Questions such as: How come melanoma appears mostly on parts of the body least exposed to sunlight, like the mid torso? Why is there such a relatively short lead time for melanoma (only a few years), whereas the less lethal skin cancers (basal and squamous-cell carcinomas) result from an exposure of up to twenty years?

In an attempt to answer that second question, doctors began to focus on sunburn rather than sunlight—thinking perhaps that fair-skinned office workers' irregular suntanning habits had something to do with the fact that it is this population most at risk. Critics say it is the use of sunscreens in the first place that stops people from developing a tan—the body's own natural sunscreen—since dark-skinned people rarely get skin cancer. There has also been speculation about the decrease in the ozone layer surrounding the earth (thereby admitting more harmful ultraviolet radiation), but ozone "holes" are thus far only occurring only in the arctic regions.

Dr. Robert Becker, in his *Cross Currents*, has a more interesting thesis about melanoma's causal factors. The highest incidence of malignant melanoma in the United States is at the Lawrence Livermore National Laboratory in Livermore, California—considerably higher than that of the surrounding population. Despite the classified nature of its military defense work, the Livermore lab is known to be involved in the design and testing of exotic weapons. The cancer epidemiology unit of the California Department of Health Services undertook a study in 1977 to investigate the excess of melanoma in those employed at Livermore, and in 1985, Drs. Peggy Reynolds and Donald Austin reported their findings. Not only was melanoma elevated above normal, but so were the rates of cancers of the salivary glands, colon, and brain (but only slightly above the population). The melanoma, however, stood out in its excess, and the researchers also noted that the cancers in general were not of the types usually caused by ionizing radiation. Some undetermined agent was causing the increases.

Unfortunately, no data were available to the California Health Service study for an assessment of exposure to the nonionizing bands. Dr. Becker had been a consultant to Livermore on the potential hazards of DC magnetic-field exposures, and he learned that some personnel were routinely exposed to DC magnetic fields as high as 1,400 gauss for an entire workday, and that the lab's standards had been set at 20,000 gauss for the body's extremities and 2,000 gauss for the body's trunk region. The human anatomy has never been exposed to such high strengths, especially for such long, continuous durations. The rise in melanoma in that particular group takes on great clinical significance, since not only were the melanoma incidences higher, but those figures were already over and above the already-mentioned increases for the normal population.

Several studies have reported significant rises in the incidence of melanoma in those who work under fluorescent lights. Fluorescent tubes have strong magnetic fields and also create a substantial exposure to UVA and UVB radiation in both the office and the home. (Such lighting, which has become ubiquitous in modern life, is discussed in more detail in Chapter 12.)

Is there something particular to our increasing global EMF environment and the rise in melanoma? Knowing what we already do about how abnormal fields can stimulate cancer cells, as well as their ability to create genetic damage at a crucial stage of cell division, many now think that such a link will eventually be made—perhaps to fluorescent lights as well as other sources, microwaves in particular.

Breast Cancer

Breast cancer is the most common cancer among women. It strikes 182,000 women annually in America alone, and of those, 46,000 will die. Over the last fifty years, the incidence of breast cancer has risen steadily, on average by 2 percent a year in most industrialized countries. In America alone, from 1973 to 1988, the rate rose 26 percent. It is now said that American women have a one-in-eight chance of developing the disease sometime during their lives, with three-quarters of the cases occurring in postmenopausal women. But more and more young women are also contracting breast cancer.

Although many of our health organizations continue to advocate early detection as a first line of defense, in fact, despite our efforts in that direction, with the fine-tuning of mammography machines and the millions who have regular scans, the mortality rate for breast cancer has remained virtually the same since the 1930s. There is apparently a statistical fluke inherent in the early-detection programs. Finding the disease earlier makes it look as if more women are surviving longer, but in reality they may only be entering the statistical pool that much sooner.

The causes of breast cancer (of which there are about thirteen forms, some more pernicious than others) are unknown, but there are some likely new suspects. One is a group of man-made chemicals called organocholines, used largely in the manufacture of polyvinyl chloride (PVC) plastics, as well as in bleaching products, disinfecting agents, dry-cleaning solutions, fireproofing, and refrigeration and in such pesticides as atrazine, DDT, and DDE (a DDT breakdown product). Chemical by-products of manufacturing such as dioxin, PCBs, and PBBs are also organocholines. These chemicals are long lasting in the environment, including within the human body. Some 177 different organocholines have been found in human body fluids and tissue. In 1992, Dr. Frank Falck, at the University of Connecticut School of Medicine, reported that in tissue samples from forty women who had undergone biopsies for breast lumps, the samples found to be cancerous had high levels of PCBs, DDT, and DDE.

It is thought that organocholines contribute to cancer in two ways: by direct mutagenic effects and by mimicking or disrupting natural hormones, especially estrogen. DDT, DDE, and PCBs are all xenoestrogens—false estrogens that bind to a cell's estrogen-receptor sites. The body does not rid itself of xenoestrogens in the same way as it does natural estrogens. Dangerous types of estrogens build up in fatty tissue like that of the breast. The higher up the food chain one goes, the more concentrated become the organocholines. Animal tissues and those of large oily fish (like bluefish) have high concentrations. Moreover, organocholines are thought to be complete carcinogens, that is, they can both create and promote cancer. This means that they may be responsible for cancers in women who are considered low risk. Still, something else is needed to set the whole process in motion, and that co-factor may be electromagnetic fields.

A possible association between breast cancer—in both women and men—and EMFs keeps coming up. As of this writing, five studies have now found an increase in breast cancer in men who are occupationally exposed to EMFs. The first study was conducted in 1989 by Dr. Genevieve Mantanowski and co-workers at the Johns Hopkins University School of Hygiene and Public Health. They found an increase in male breast cancer among young telephone-company workers. Since then, four other studies have found similar increases in male breast cancer among those occupationally exposed to EMFs. In 1992, Dr. Dana Loomis, at the University of North Carolina, found a doubling of breast cancer deaths among male electrical workers under the age of sixty-five. Before that, Dr. Paul Demers and colleagues, at the Fred Hutchinson Cancer Research Center in Seattle, reported a sixfold risk increase in some younger electrical workers, and Drs. Tore Tynes and Aage Anderson, of the Cancer Registry of Norway in Oslo, reported a doubling of risks in electrical-transport workers such as train operators.

Breast cancer in men is extremely rare, and any such increase in a select population with a specific occupational exposure to EMFs has important implications for the general female population. The physiological link between EMFs and breast cancer (as well as other glandular cancers, like prostate cancer and lymphoma) may be through the suppression of melatonin produced by the pineal gland in the brain.

Several studies now find a correlation between low melatonin levels and breast cancer, as well as suppression of the immune system itself. A handful of residential surveys and occupational studies are also showing associations between female breast cancer and EMFs. In 1991, a study conducted by Dr. John Vena and co-workers at the State University of New York at Buffalo found a small increased risk of breast cancer in postmenopausal women who regularly used electric blankets throughout the night for at least ten years, although the findings were not considered statistically significant.

The 1982 epidemiological studies of Dr. Nancy Wertheimer and Ed Leeper found a threefold increased incidence of breast cancer in women under age fifty-five living near high-current power lines. Dr. Dana Loomis found a 40 percent higher mortality rate from breast cancer in women employed in traditionally all-male electrical occupations. The suppression of melatonin at night (when levels are supposed to be elevated) from exposure to electric blankets, domestic wiring, appliances, or just simply light itself after dark, may be contributing to the steady rise in breast cancer rates. Dr. Richard Stevens, at the Battelle Pacific Northwest Labs in Richland, Washington, was the first to theorize that nighttime melatonin suppression might be a contributing factor in breast cancer incidence.

There are likely synergistic actions between melatonin suppression and oncogenic cells or other carcinogens. In 1992, Dr. Margrit Wiesendanger, of the Lawrence Berkeley Laboratory in Berkeley, California, reported that extremely low frequency (ELF) magnetic fields blocked melatonin's ability to control the growth of human breast cancer cells in laboratory samples. Prior to that, she had verified the work of Dr. David Blask, of the Mary Imogene Bassett Hospital Research Institute in Cooperstown, New York, who had found that the growth of breast cancer cells in lab samples could be inhibited by melatonin. And Drs. Robert Liburdy and Paul Yaswen, also at Lawrence Berkeley Laboratory, found that melatonin's protective effects were thwarted by

60-hertz magnetic fields as well. It is thought that some magnetic fields may influence the way melatonin binds to the cell surface or that magnetic fields enhance cancerous cell growth directly.

Animal studies also found a power-frequency and breast-cancer connection, in the work of Dr. Wolfgang Loscher's research group at the School of Veterinary Medicine in Hannover, Germany. They reported that the growth of breast tumors in rats was promoted by EMF exposure. And recent research designed to simulate exposures from the Maglev transportation systems now in operation in Europe found a significant decrease in an enzyme called seratonin-N-acetyltransferase (needed to control melatonin production in the pineal gland) in rats exposed to certain intermittent fields.

Are we seeing some combination of exposure to organocholines (or other agents) in the environment, acting as abnormal estrogens, in combination with the suppression of melatonin due to ubiquitous EMF exposure, working together in a detrimental way? This may well prove to be so, but getting a definitive answer will be difficult if we try to use the traditional study design of exposed groups and nonexposed control groups. There may be no such thing as a nonexposed EMF population today. Nevertheless, several EMF-and-breast-cancer studies are currently being done. In the meantime, breast cancer clusters among VDT users are beginning to be reported.

PART THREE

WHERE THE EXPOSURES ARE, HOW TO MEASURE, HOW TO MITIGATE

Chapter 11

GETTING A HANDLE
ON OUR ALTERED ENVIRONMENT

What Have We Done?

T HERE IS SO MUCH artificially generated light on earth today
that a person can literally read outside at night in almost any
major city. By anyone's reckoning, that's an altered environment.

Many people who are born and raised in urban areas never
get a real glimpse of natural starlight or experience the backlit
silver luminescence of the winter landscape when there is snow
on the ground and a full moon in a clear night sky. Instead, whole
generations now experience night light as the grotesque red-
orange glow of sodium-vapor streetlights—the same color that
used to alert distant neighbors that a huge fire had broken out.

A mere fifty years ago, this was not so. Everyone, even the
sophisticated urbanites of their day, experienced nighttime not as
something to conquer and subdue but as a mysterious natural
complement of the busy daylight hours. Today, astronomers are
having increasing difficulty finding a dark place on the planet
from which to view the night sky. And this change in the electromagnetic environment represents just the light frequencies.

The earth today is literally blanketed with a range of artificially generated frequencies throughout the electromagnetic

spectrum. The radio/microwave frequencies have proliferated at an alarming rate and continue to do so, nearly unchallenged. And the gigahertz ranges may well do the same within the next decade.

This was a cause of concern as far back as 1971. A White House advisory committee called the Electromagnetic Radiation Management Advisory Council said that nonionizing radiation from radar, television transmitters, communications systems, microwave ovens, industrial heat-treatment systems, medical diathermy equipment, and many other sources was permeating the modern environment. In a call for a national research effort on the risks of radio-frequency radiation, the council went on to say that since 1940 the growth of such radiation sources had been "phenomenal," and that there was increasing anxiety, even at low power densities, that such sources could adversely affect biological organisms. And this was long before cellular phones, cellular towers, personal computers, MRIs, ultrasound technologies, radar guns, satellite uplinks, pagers, and a host of consumer products like wireless audio speakers and over 100-ampere domestic electric service became synonymous with our high-tech lifestyle. Once upon a time, an office consisted of a typewriter and a dial phone positioned on each desk. Now it is eight hours of environmental exposure to cross-frequencies generated by copiers, computers, fluorescent lights, numerous kinds of screens, faxes, modems, and printers.

Nowhere on earth can one escape these exposures altogether, although some places have far fewer strengths than others. In fact, if you live in an area with poor radio or TV reception (without a cable hookup), consider yourself lucky from an EMF-exposure point of view unless the reason for bad reception is interference from other nearby transmissions.

Many exposures are global in scope, even in rural environments where no communications towers or high-tension lines are visible. For instance, before its breakup, the Soviet Union had a massive over-the-horizon radio transmitter, nicknamed Woodpecker, that transmitted radio signals modulated at around 10 hertz, the frequency we may be most biologically attuned to. (Brain waves during sleep are in the 10-hertz range.) The signals could be heard all over the world, which meant that, wherever you were, you were being continuously dosed with an artificial 10-hertz pulsed wave from that one source alone. Astronauts can see the electrical imprint from high-tension lines all the way into space. The U.S. military planned an enormous 6,000-square-mile underground transmission grid (originally called Project Sanguine) in Michigan capable of communicating with U.S. submarines anywhere on earth, at any depth. Communications satellites create a spray of frequencies directed toward the earth, which is what we "receive" in backyard dishes. And the U.S. military is experimenting with powerful earth-stationed RF transmitters that would alter the ionosphere, in effect creating a virtual antenna in the ionosphere to communicate with submerged submarines. These are to name but a few of the military and civilian sources of EMFs.

It is long past time for us to question the intelligence of all this. Instead, we rush headlong to develop new electronic gadgets, virtual-reality headgear, seamless cellular-phone capabilities stretching from coast to coast, and information superhighways. The financial pages of the world's best newspapers constantly tout revolutionary devices like motion detectors mounted on every car, instant satellite information services to replace road maps and atlases, and wireless laptop computers that can send data anywhere, anytime—with no mention whatsoever that there may be health risks associated with them. Perhaps financial reporters do not know there are health concerns, but they should. Investors and the general public will presume such products are safe unless responsible journalists learn to tell them otherwise.

We must ultimately put many of the underlying questions to ourselves—the consumers. Do we really need another radio station every sixteenth of an inch on the dial, with its accompanying transmission facilities adding to this abnormal blanket? Do we really have to be available every second of the day via cordless telephone, even when walking down the street or riding in a car? Do we really need to wear little television sets on our wrists? Beepers on our belt loops? Do we need to bombard ourselves in our homes with radio frequencies from a new generation of wireless products?

Cordless phones, which broadcast a weak FM signal to the base unit from approximately fifty feet away, may also couple with the domestic phone wires and effectively turn the entire house into a radio-frequency wave in some circumstances. And this is not to mention the fact that *many* of these frequencies overlap into odd EMF couplings and amplifications that were never intended, have never been studied, and were not even vaguely anticipated when the few inadequate standards that do exist were implemented. This is true of domestic exposures as well as public works projects. For example, placing cellular-phone towers near high-tension lines means that the lines can act as corridors along which the microwave frequencies from the phone towers can travel. This creates whole new exposures for anyone living or working near the high-tension lines, yet the issue of added frequencies never comes up in siting considerations—perhaps because such projects are so unpopular in local communities that it is politically easier to group them together.

Another aspect regarding our total exposure that is only beginning to be talked about, although not widely, has to do with the *additional* cumulative amount of EMFs from many different sources, including domestic appliances and hand-held devices like hair dryers and power tools—anything we touch or stand near that has high magnetic fields. As briefly mentioned in Chapter 4, anytime you touch a conductive object, like a refrigerator with current flowing through it (which creates magnetic fields), contact currents can be induced in your body at the same frequency. While such currents are thought to be fairly superficial, in fact their repeated inducement may have consequences, especially when factored in over a whole twenty-four-hour period. Magnetic fields have deep penetration ability and can induce fields of their own. Intermittent exposures throughout the day means a constant state of internal flux.

The episodic or intermittent exposures to magnetic fields of sufficient flux density (such as standing near the refrigerator off and on all day, or a hairdresser using a hair dryer throughout the day on customers) can cause measurable physiological changes. Such changes have been repeatedly demonstrated in laboratory animals and in human studies. Effects on the human cardiac interbeat and on brain-wave patterns have been more pronounced with intermittent magnetic-field exposures than with steady-state magnetic exposures at the same intensity. In experiments with hamsters, fifteen minutes of exposure to 60-hertz magnetic fields at 1 gauss delayed by several hours the normal rise in nighttime melatonin production. Exposing rats to short-term pulsed DC magnetic fields has had similar results. And again, these are only a few of the studies.

The popular new digital technology, which will soon be applied to just about everything in today's high-tech world, functions through a kind of on/off signal that in effect creates a pulsed exposure. We may be heading in the wrong direction with digital equipment, given what we already know about pulsed versus steady-state magnetic fields.

In the past, such appliance-type exposures were considered transient and of little consequence, but this may prove untrue. Not only do chronic, low-level, long-term exposures from various sources such as close proximity to electrical lines and transmission towers need to be considered, but it appears that intermittent ones from consumer products do as well. In other words, we may need to reduce all of our exposures as much as possible. It is called prudent avoidance, and there are many ways to approach it.

Prudent Avoidance

Prudent avoidance is exactly what the term says: reduce your EMF exposure as much as you can without inconveniencing yourself unduly or unsafely. No one, for example, would recommend doing without refrigeration, since the danger of food contamination is life threatening. But don't sleep near a refrigerator, and try to arrange your kitchen so that the work space where you spend a majority of time is at least three feet from any operating appliance or machine. This includes under-the-cabinet installations for microwave ovens and a host of small appliances. Turn off appliances that are not in use, including computers, and sit several feet away from a computer monitor and at least six feet away from a television set (although some people think that three feet is enough). If you use a computer with a CRT monitor (as opposed to a liquid-crystal display), try to use a keyboard that detaches, so you can get farther away from the monitor.

Be conscious of high-exposure areas, such as the kitchen. Microwave ovens produce the strongest EMFs — sometimes as high as 10 milligauss when turned *off*. Usually you need to be five feet away from a microwave oven to achieve 1 milligauss or less. Don't allow children to play in the kitchen while you are involved with meal preparation. Delay running appliances such as the dishwasher until you are away from the kitchen. If your ovens are electric, try to use them when you can be out of the area. Or switch to gas appliances. Keep as far away from operating appliances as possible. Use battery-powered ones, and try to keep the speed settings on appliances low.

Perhaps return to nonelectric carving knives, razors, toothbrushes, screwdrivers, and the like. Use a cordless phone only to answer when you are outside, but come indoors to continue the call on a corded phone. Restrict car phone and cellular phone use to emergencies only. Change fluorescent lights to incandescent ones. Position motorized electric clocks at least three feet away from your head at night. Do not use an electric blanket other than to warm the bed before getting in—make sure it's unplugged, not just turned off. Rethink using waterbeds. Investigate where the electricity comes into your home. Try to rearrange sleeping/working space so as to be as far away as possible from outside electrical hookups.

Insist that manufacturers make low-EMF-emitting devices. Request that *Consumer Reports* measure and publish EMF emissions as part of their standard review of appliances and devices. There are three different bills making their way through Congress, as of this writing, that would require manufacturers of consumer devices to label their products for peak EMF emissions. Contact your representatives and tell them you support such legislation. (But be aware that it will take time for manufacturers to redesign some of their products.) Also, be supportive of utility companies trying to mitigate exposures by buying larger rights-of-way or burying or redesigning their wire configurations to cancel fields. Some of these changes may turn up as higher utility rates.

When designing living space, stay away from electric baseboard heat or ceiling-cable configurations. Position electric space heaters away from where you sit or sleep. If you do have electric baseboard heat, be sure that beds and cribs are at least three feet away from it.

Prudent avoidance has also become part of a broader consciousness. Several states, Colorado and Wisconsin among them, now require that utility companies consider the health effects of power-line exposure when planning new transmission lines. Corporate prudent avoidance is defined as striking a reasonable balance between the potential health effects of exposure to electromagnetic fields and the cost and impact of mitigating such exposure. It means that state public-utility control commissions can require utilities to use the best available technology to reduce emissions through new wire designs, and to site new lines and substations away from schools, day-care centers, hospitals, and the like. (One day, this may apply to neighborhood distribution lines, too.)

Some states think prudent avoidance is too radical, in the face of the scientific uncertainties, and they recommend what they call voluntary exposure control. This entails the providing of information to concerned individuals, but it makes no official recommendations and does not address corporate responsibility. Critics say it is a hollow public-relations gesture.

Some European countries are beginning to make magnetic-field exposure recommendations for their civilian populations. Sweden is in the forefront, with a draft proposal for standards at 2, 5, or 10 milligauss for new residential buildings near existing power lines and substations; recommendations for larger rights-of-way are also expected. Swedish labor unions have called for an average 2-milligauss limit for workers. Unfortunately, even in progressive Sweden, these proposals have been stalled.

Specific prudent-avoidance recommendations will be made throughout the subsequent chapters.

How to Measure Electromagnetic Fields

There are several ways to measure the electromagnetic fields around you. The easiest and cheapest way to "experience" the electric fields alone is with a regular battery-operated AM transistor radio. Tune it in between stations, so you hear only "white noise," and then walk around and listen to the changes in static at various EMF sources. Transistor radios are particularly sensitive to electric fields, which is what you will be hearing, not the magnetic fields. Although it may be possible to "surmise" the magnetic fields, which accompany some of the white noise around TVs and computers, if you want to "listen" to 60-hertz magnetic fields, an inexpensive telephone amplifier (often used by the hearing impaired) that fits over the receiver will allow you to do that. (Radio Shack makes one for $10.95.) You do not have to connect it to the phone. It is battery powered, and you just hold it up to your ear and walk around.

Some of the things you will hear with the radio or the telephone amplifier will be basic static at various strengths near certain appliances; much stronger sounds with several different types of tone near a computer; almost the whole multifrequency band near a TV; another kind of tone near a light dimmer; and still another range of tones near an ultrasonic humidifier. One of the most startling demonstrations is to stand near a TV and turn it on with the transistor in hand. (Do it with the volume on the TV off, so that all you hear is the radio static.)

The radio is sensitive to RF fields but not to 60-hertz fields and will give you only a general idea of the different sources, how to locate them, and at what distance the fields begin to dissipate. With the TV turned on, back away to the point at which the static noise disappears. This is approximately a 1-milligauss range. You will have to turn the radio several different ways because antennas are directional. Find the quietest place in the room. Although this method is imprecise, it is a good way of figuring out where to position the furniture.

Wearers of hearing aids have a built-in RF detection device—although such background noise is often a nuisance for them.

For more precise measurements of magnetic fields, especially in the 60-hertz range, you might want to either purchase or rent a gaussmeter. Good inexpensive models are available for around $200, and many of the companies that sell such equipment also rent it. And an increasing number of electrical-engineering companies will take measurements for a fee. Check your phone book or newspaper to find such firms.

Also, local utilities will often come and take measurements for homeowners (although not always for tenants) if requested to do so, especially if you live near a high-tension line. Usually they don't charge anything. They will sometimes make mitigation recommendations if they find a "hot spot" in domestic wiring, for instance, but typically they will just measure different areas of the house or neighborhood without necessarily interpreting the numbers for you.

The utility company can refer you to an independent firm in your area to take measurements, if you prefer, but the advantages of working with the utility company first are that they often take the time to educate you, plus they have trained technicians and better broad-band equipment than most nonprofessionals can afford. The advantages of working with an independent agent are that a conflict of interest is less likely and you will get mitigation recommendations as well as a written report afterward. Hiring independent firms can be expensive, but it is helpful to have someone put everything into perspective for you. For the $300 or so that such firms usually charge, however, you can buy your own gaussmeter.

In setting up an appointment with someone, be sure that it is during hours of peak electrical use, in order to get an accurate assessment of your highest exposures. Measurements made at mid morning or late afternoon will not tell you what transpires at dinnertime, when there is high electrical use, or at noon, when the baby is taking a nap.

Firms that take EMF measurements report that increasing numbers of prospective home buyers obtain such information before making a decision. EMF readings are becoming as important as radon tests in environmental assessments.

Gaussmeters

There are several different types of gaussmeters for sale today, ranging in price from around $100 to over $6,000. The most expensive models are really only needed by those in the industry and will not tell the consumer much more valuable information than the less costly models. This is not a case where expensive is automatically better. There are some good models available for around $200.

Some companies sell separate sensors that can attach to

simple voltmeters sold in any hardware store or electrical-supply house. But sometimes their sensors are specifically calibrated for the meter the company sells. Ask if this is the case or if sensors can be used with other meters. If you already own a voltmeter, ask if their sensor will attach to it and if it will read accurately.

Other models come as one unit and can be quite compact. They are often the size of a cigarette pack with a large digital readout screen. Whichever meter design turns out to be the most convenient model for you will depend on what you are looking to measure. The one-unit models are the easiest to carry around and will fit inside a purse or jacket pocket. You can take measurements anytime, anywhere—and they are especially handy when making a new appliance or equipment purchase. You can measure various items in the store before bringing them home.

But a meter with a separate sensor allows you to get readings from difficult-to-reach areas, such as underneath a waterbed or behind a computer, where the single-unit model would be useless. The two-unit model also allows you to make measurements at different heights, by holding the sensor high while keeping the readout unit within easy viewing distance. On two-unit models, you can also use the voltmeter separately, with a different probe, to see if wall outlets and batteries are live.

Sensors contain small coils and typically come in 1-axis or 3-axis models. The 1-axis models are good for locating the direction from which exposures are coming, as well as the strength of specific fields. For the general consumer, they will give the most information for the least money. The more expensive 3-axis models simultaneously measure fields in space from several directions. Often they contain small computers that make computations from three different directions and do the necessary mathematical conversions to come up with a single number. They are more appropriate for the professional who takes many measurements in a day and needs a built-in memory chip. Also, 3-axis models will not give you specific information about individual appliances, which is what the consumer needs to know. If you keep rotating a 1-axis model, you can still get spatial measurements away from EMF sources. The highest number always prevails.

Gaussmeters measure power-frequency fields with varying degrees of accuracy. They will tell you nothing about RF exposures in your neighborhood, although some very expensive models can measure broad-band frequencies all the way into the microwave range. Such an RF-measuring device does not exist yet in an affordable model for the nonprofessional. If you live near a radio tower or want to measure exposures other than 60 hertz, a gaussmeter is not for you. Ask an electrical engineering firm for assistance.

How to Read a Gaussmeter

Gaussmeters can be a bit frustrating at first, not because they are difficult to use but because they may be confusing to interpret.

Gaussmeters measure *magnetic* fields. As we explained earlier, the basic units of measurement are expressed in gauss and in tesla, both of which measure the same thing—not unlike feet and miles, which both measure distance. Many laypeople are intimidated by all this, but in truth it is no different than measuring, say, weight in ounces or pounds.

Electric fields are measured in volts per meter (V/m) or, when the field is strong, in units of a thousand volts per meter (kilovolts, or kV/m). Magnetic fields are expressed in gauss and tesla, which are both large measurements, so milligauss (mG) are often used. A milligauss is one-thousandth of a gauss; and there are 10,000 gauss in 1 tesla. Sometimes you might also see microtesla or nanotesla. A microtesla (μT) is 10 milligauss; a nanotesla (nT) is one-tenth of a milligauss. Milligauss and gauss have become the most commonly used terms. The following table lists some of the equivalent numbers, showing the decimals as they might appear on a meter.

Most gaussmeters measure both milligauss and gauss with

1 tesla (T)	= 10,000.00 gauss (G)
1 gauss (G)	= 1000.00 milligauss (mG)
1 milligauss (mG)	= 0.001 gauss (G)
1 microtesla (μT)	= 10.00 milligauss (mG)
1 nanotesla (nT)	= 0.010 milligauss (mG)

the flip of a switch, and most also measure a tenth of a milligauss. Unfortunately, meters are not uniform from manufacturer to manufacturer in their readout screens, and it may take a little time to figure out how to read yours. One milligauss on some digital meters reads 001. On most it reads 1.0 when the meter is switched to the milligauss setting. On those meters that also measure a tenth of a milligauss, it will look like 0.1. (For those wanting extremely accurate assessments, those tenths of a milligauss can add up and become important in the measurements.) Anything reaching the gauss range, which is a thousand times higher, or 1,000 milligauss, will look like 1.00 and up when the meter is switched to the gauss setting.

It is best to move slowly when using a gaussmeter and rotate it often in several directions. Be careful around power lines and always be aware of your surroundings, as you may be in a dangerous area. Sometimes the numbers on the meter change radically within inches of the last measurement. And sometimes—if the batteries are going dead, or if you are moving too fast, or if you are in the presence of a strong RF field that the meter is not shielded against—they may fluctuate wildly. Under normal circumstances, however, the highest number you get while standing still is the accurate one.

When you first get a meter, play with it to get a sense of what is around you. Measure all over the surface of individual appliances, turned on and turned off. You will quickly discover that the highest fields are near the motors and control panels. Measure space in between different appliances when they are switched on to measure their EMF interactions. In a high-use room like the bathroom or kitchen, make measurements in a real-life scenario in which everything is turned on at the same time. Make room-by-room measurements. Pay attention to the wall on the other side of an appliance when it is on; magnetic fields can easily penetrate walls. Is there a bed opposite a TV in the next room? A chair opposite a computer? Measure dimmer switches and switches that control one light from two locations. Measure both sides of hallways at various heights where wiring might run behind the walls. Measure floors. Measure the basement and the electrical boxes. Write the measurements down in a log. Make sure to note the time of day that you take measurements, since they can fluctuate throughout the day depending on peak electrical loads. Measure three different times during the day. Pay special attention to areas where you or your children spend a lot of time, especially bedrooms. Measure all around the area where your head is placed in bed, as well as about eight inches above the bed itself.

Later, make measurements outside, all along the perimeter of your property and especially where the electrical wires come into the house. Be careful when doing this not to touch a live conductor and form a ground for the current with your body.

A gaussmeter will give you an excellent sense of how sharply even some of the strongest 60-hertz fields drop off with distance from an appliance. The general rule of thumb for a safe distance is about three feet from any generating source. Practice the habit of switching on an appliance and stepping back from it by a yard, instead of standing over the stove waiting for something to boil or lingering over the toaster waiting for it to pop. You can measure the length of your arm and use your arm's length as a guide. Some parents mark a six- or eight-foot distance from the TVs with tape on the floor, to keep the children back, but even a three-foot stay-behind line will help.

When measuring computers or other office equipment, measure the equipment itself, and then slowly move the gaussmeter back to where you normally sit or stand. Measure at the level of your head, neck, chest, and groin. (This is where a two-unit model comes in handy.) Try to measure between various operating machines to determine cross-fields. And remember to measure any walls in common with another office or an elevator shaft.

What Is a Safe Number?

Unfortunately, a true "safe" magnetic-field exposure as of this writing is still unclear. In fairness, this is largely due to the inherent complexities of the subject. Researchers are still debating what they are supposed to be measuring, because no one knows what the detrimental element in the electromagnetic equation is. It is presumed at the moment to be the magnetic component, since there is more evidence pointing in that direction, but even within that presumption questions arise. For instance, are there particular wave forms that are more detrimental? Are the rounded sine waves (typical by-products of electrical transmission) all right at one level of intensity, but the jagged sawtooth waves (typical of TVs and VDTs) hazardous at the same intensity?

What about harmonics—the phenomenon of additional frequencies at 120 and 180 hertz that are also created by devices and appliances that travel out from wires just like 60-hertz waves? Sometimes the harmonic wave is more dominant than the 60-hertz wave. Some of the better gaussmeters measure harmonics, but with varying accuracy. (The issue of harmonics may be an important one in the future because the 180-hertz wave can actually build up on some wires, especially around computers.)

For the consumer wanting a simple answer on safety, these kinds of issues can seem like techno overkill, but they are genuine concerns to those involved with fine-tuning our knowledge. Office environments and fluorescent lights, for instance, are rich with harmonics. Gaussmeters that do not factor them in will not provide an accurate assessment of exposures. (Ask the manufacturer if the meter you are buying measures harmonics, but don't be surprised if you get a complicated or evasive answer. The question itself is more complicated than can be explained here.) EMF office environments can be extremely complex. It is probably best to hire an independent firm with broad-band equipment to do the measurements. A simple consumer's gaussmeter that may not read harmonics will not give you accurate information.

While no one wants to make a recommendation that turns out to be incorrect, in general those who have examined the medical literature recommend trying to minimize exposures to around the 1-milligauss level—knowing full well that even this may turn out to be too high. Or that a recommended low level may be found to fall within a dangerous window for bioeffects that a higher level does not. This is still evolving knowledge, but the 1-milligauss level is where the least bioeffects have been observed; 2 milligauss and up are the levels at which various detrimental associations begin to appear.

Some would say the 1-milligauss recommendation is too extreme. Others would call it a conservative educated guess. But maintaining a 1-milligauss level is not always easy, or even possible, as we move through a normal day. The ambient EMF background of modern cities, according to some estimates, is around 3 milligauss, although this varies from area to area. (Some streets measure 0 milligauss, others are 100, depending on plumbing currents, unbalanced loads, and peak electrical usage.) Moreover, the exposures increase radically the higher up in altitude one gets on most city skyscrapers. That is why it may be important to reduce as many exposures as possible from extraneous consumer products.

Anyone living or working in an ambient environment with consistent readings above 5 milligauss will certainly want to investigate how to reduce those exposures.

Some Gaussmeter Models
In response to the rising demand, twenty-five companies now offer over sixty different gaussmeter models, some of them more

accurate than others. The only existing accuracy standard is that of the National Institute of Standards and Technology of the Institute of Electrical and Electronics Engineers (IEEE), but compliance with it is voluntary. Called the ANSI/IEEE Standard 644-1987, it specifies that meters must be within 5 percent, plus or minus, of the correct reading.

The Environmental Protection Agency (EPA) tested a number of models in 1992 and found wide differences in accuracy. There was criticism of the EPA's test approach, however. Some said there was an absence of uniformity in the way different models were tested, so comparisons were unreliable. In addition, some of the manufacturers with low ratings have since improved their models, and not all the models tested are still available, or they have evolved into newer versions. In other words, the EPA survey is no longer current and may not have been accurate in the first place. We can hope the EPA will tighten its test protocols and regularly publish updated data on gaussmeters that will be useful to consumers.

Gaussmeters can measure different electromagnetic properties and be more accurate in some readings than others. For instance, one meter may be accurate at the 60-hertz frequency but be off by a huge percentage at the 180-hertz harmonic. Some are better with sine waves than sawtooth forms. But these distinctions are more technical than most consumers care to get. Accuracy data continue to accumulate, and manufacturers are improving their models all the time. As of this writing, the following table lists the top five companies that make reasonably accurate consumer models according to the IEEE standards for 60-hertz sine waves. (All are equally good; they are listed alphabetically.)

Although there are cheaper models available, the two best values for accuracy, measurement of harmonics, and two-unit convenience appear to be the MSI-25 and the EFM-131. These two models are also less likely to drift out of calibration and lose their accuracy over time (something that plagues other models)

Electric Field Measurements 86 Interlaken Rd. West Stockbridge, MA 01266 Contact: Dr. Don Deno (413) 637-1929	EFM 140 (sensor only) EFM 131 (sensor and meter)	$ 75.00 115.00
ExpanTest, Inc. 232 St. John St., Suite 132 Portland, ME 04102 Contact: Robert Wagner (207) 871-0224	ELF Sense (Model 1A)	340.00
Holaday Industries, Inc. 14825 Martin Dr. Eden Prairie, MN 55344 Contact: Dave Baron (612) 934-4920	HI-3604 (ELF)	1,295.00
Magnetic Sciences International HCR-2, Box 850-295 Tucson, AZ 85735 Contact: Karl Riley (800) 749-9873 or (520) 822-1640	MAG CHECK (sensor only) MSI-25 (sensor and meter)	92.00 215.00
Meda, Inc. 485 Spring Park Pl., Suite 350 Herndon, VA 22070 Contact: Barbara Vayda (703) 471-1445	PLM-100	395.00

because their separate probes do not have to be amplified by electronics. In addition, the face of the MSI-25 voltmeter has been altered to read like a true gaussmeter, for the consumer's convenience. The settings on most standard voltmeters can be extremely confusing to the layman.

Other manufacturers will probably have new models out soon. For a current list of gaussmeters, models and prices, send $1.00 and a self-addressed stamped envelope to *Microwave News*, P.O. Box 1799, Grand Central Station, New York, NY 10163. *Microwave News*, the leading newsletter on the subject of EMFs for nearly two decades, keeps a list of manufacturers, but it has not independently tested the models, nor does it make purchasing recommendations. Some models are listed for as low as $39.95, but call the company and ask for details before you buy. You may not get what you hope for.

Many manufacturers have 800 numbers to call for information. Ask which models are made for the nonprofessional, how high in gauss they measure, and if the manufacturing and accuracy standards match those recommended by the IEEE. Also be sure to ask manufacturers if they have support material for the nonprofessional. Some of the information that comes with gaussmeters is written in language that is too technical for the average person. The instructions sometimes contain mathematical formulas. There is wide variation between manufacturers regarding written material, perhaps because they are engineering firms. Those that develop a line of products and support material for the average consumer will probably do the biggest business in time. But this aspect of the business is also still evolving.

Magnetic Sciences International, the maker of the MSI-25, puts out an excellent thirty-three-page booklet to go with its gaussmeter, which explains everything for the consumer in accessible language, including how to use and read the meter. Also, the EPA has reopened a toll-free information number in order to relieve its regional offices of some of the burden of fielding

consumer questions about electromagnetic fields. The number is
800-EMF-2383; call between 9:00 A.M. and 5:00 P.M. eastern
time. You will get middle-of-the road information about what is
known and not known. It may have information about gaussmeters as well, but don't trust that out-of-date 1992 survey.

Chapter 12

WHAT'S AT HOME?

L OOK AROUND YOU. If you live in a typical modern home, you are surrounded by electromagnetic fields from numerous sources, beginning with your home wiring. In America, electrical power runs at 60 hertz. In Europe, it is 50 hertz. Some say that this difference of a mere 10 hertz is a significant one, that there are more bioeffects at 60 hertz than at 50. Sixty hertz may in fact turn out to be an important window for human bioeffects. But we don't know yet.

Today's modern home, however, is more complex than just the 60-hertz frequency. Light-dimming switches, cordless phones, wireless technologies like remote-control devices for VCRs, TVs, and garage-door openers, and wireless audio speakers all use the higher radio-frequency bands. Compact-disk players use the laser/light frequencies; TVs and CRT computer screens use a broad-band range that begins at 60 hertz and may go all the way into the ionizing bands around the X-ray range.

In addition, appliances that were formerly 60 hertz only, such as electric stoves, are now coming in high-tech models using halogen heating elements or featuring continuous-control dials, which are in the light and RF frequencies, respectively.

One interesting fact about our modern environment, through, is that in some ways the newer technologies actually create *weaker* electromagnetic fields than many of the older appliances and former methods of domestic wiring. Analog electric clocks are a good example. They are the old-fashioned kind, with a motor that moves the hands around the dial face. Anything motorized gives off a significantly higher electromagnetic field than something with digitally lighted numbers, and battery-operated models give off none, because batteries are direct current, not alternating current. (Battery chargers, however, give off significant fields.) And so, in this one respect, modern stoves with their lighted numbers are lower in EMFs, although some of the quick-heating coil models have high fields even without analog clocks. This difference in EMFs between analog and lighted digital models would also be true of clocks and clock radios.

Also, the old way of wiring houses sometimes separated the "hot" wire from the neutral, which caused much higher magnetic fields. Today's method of wiring positions the lines side by side, and this closer proximity helps cancel out the fields or greatly reduces them. So that is the good news. The bad news is that this reduction was a by-product of making the system more efficient and had nothing to do with intentionally trying to lower EMFs.

No major manufacturer of domestic appliances (with the exception of electric blankets and computers) as of this writing designs low-EMF products. If you call a customer representative of any major manufacturer and ask about the EMF readings for a new appliance you are thinking of buying, the representative probably won't be able to give you an answer, and might not even understand the question. If you ask what the EMF difference is between an electric stove of the standard coil style, one with a radiant/halogen smooth top, and one with an induction smooth top, the customer representative more than likely won't know. In the best companies, someone will find out and get back to you. The rest will adopt a public-relations stance and try to assure you that all of them are safe. If you want a solid answer about numbers and emissions, try to get to someone in the engineering department. There, unless they have all been instructed not to talk to consumers, you might get an answer.

You might also get a gaussmeter and talk a store salesperson into hooking up an appliance so you can take your own readings, but this could be easier asked than accomplished. Most store models do not have the electrical cords attached or the proper 240-volt outlets available. And many people purchase an appliance without ever seeing the actual model. So phoning the manufacturer might be an important avenue of information.

Best advice: be persistent. Tell the manufacturer that you will return the item if, after you measure it yourself, it has higher fields than what you were told. And say the same thing to the salesperson, so that everyone is clear about your intentions.

How Fields Can Cancel Out

Electromagnetic fields can add to and subtract from each other, thereby either strengthening or weakening themselves. It is one factor being explored by power companies as a way of reducing magnetic fields along rights-of-way. And it is something that manufacturers should be encouraged to explore in creating low-EMF products. Often, it's only a matter of simple rewiring or the use of shielding materials that would cost pennies for most appliances.

The way fields cancel each other works like this: if you had two separate 60-hertz electric fields at the same place and they were exactly in phase with each other, meaning that they are alternating in strength and direction together, then their individual strengths would add together. (Two similar 6-ampere conductors, each putting out 12 milligauss at 1 meter, will together put out 24 milligauss at 1 meter.) But if the two fields are exactly out of phase, meaning that one reaches its greatest strength in one direction precisely as the other achieves it in the reverse direction (called reverse phasing), the fields would cancel and the value would be 0 volts per meter. (The illustration entitled "Rephasing," on page 340, shows how this works.)

One mitigation approach being considered by power companies is the alteration of their phasing to reduce fields through partial cancellation. It would save them millions in rewiring schemes. Several other possible approaches for utilities are discussed in more detail in Chapter 14.

Some electric coil-style stoves have actually achieved canceled fields by accident. Manufacturers need to ask their engineering departments to put their minds to achieving the same on purpose.

Domestic Wiring*

A substantial amount of abnormally high magnetic fields in the home (and in the workplace) comes from incorrect wiring practices, which can be detected and corrected. Sometimes abnormal fields result from common grounding configurations, and there are ways to fix them that will still honor building codes.

A professional mitigation consultant should be able not only to identify the source of such high fields (anything consistently above 3 milligauss), but also to direct the electrician (since often the cause is the presently required practice of grounding to water pipes) on how to mitigate for them.

A note of caution: grounding is essential for safety; otherwise, every time you touched something conductive that had a hot wire "shorted" to it leading, for example, to a metal faucet or appliance, your body could complete the circuit and you could receive a potentially lethal shock. There are numerous cases of people being zapped while touching a defective toaster with no ground wire and the kitchen faucet at the same time.

*The following sections were adapted with permission from *Tracing EMFs in Building Wiring and Grounding* by Karl Riley.

If you discover that your well pump or some other domestic system has a wire attaching it to a pipe (called a bond), do *not* disconnect it without the advice of a professional licensed electrician, or your local building inspector. The National Electric Safety Code mandates certain grounding regulations to prevent fires and electrical shock.

Laypeople often find the language in the electrical profession frustrating and difficult to grasp, especially the use of the term "ground." Nonprofessionals think of ground as "the ground," meaning dirt or the earth. Electricians use it generically, as in "a ground," which is something that provides a conduit or path along which electricity can travel. Then they speak of "grounding," which is the process itself. It may help to think of the system as a road map. Electricity travels. Grounding paths are a sort of highway designed to preserve the order of things.

A grounding electrode (called a "ground") serves two purposes: to help dissipate into the earth excess electricity from a lightning strike or an accidental power surge (as when a tree falls on a power line) and to help keep the entire electrical system steady at 120 volts—otherwise the voltage could drift up over 120 volts-to-ground and create a hazardous condition.

There are many misconceptions about the role of a local grounding connector to the earth (this is the metallic wire that runs from your electric meter into the earth or out through underground pipes in your neighborhood), especially among electricians. It is widely presumed that this ground is to clear a fault current (something unexpected that occurs within the domestic electrical system), but actually that situation is handled by the utility company's neutral return lines back to the transformer (which will be discussed shortly). In truth, the local ground is to handle unexpected high power surges and to stabilize the entire system, not to channel fault current. It is meant to persuade extremely high voltages into a safer place, rather than have them rumble through and destroy the network.

There are differences, too, between aboveground paths

within the domestic system, such as metal water pipes, air ducts, and construction steel, and paths below ground, which include outside water pipes and telephone or cable ground sheaths. At one time, gas pipes were also used, but this practice was outlawed in the early 1990s.

If you live in a single-family home with a single service drop or private well system, locating the grounding configuration will be easier. If you live in a suburban neighborhood with town water supplied to several houses along an interconnecting pipe system, or in a high-rise apartment, it will be more difficult.

Elevated Magnetic Fields in the Home

These are the factors that will elevate magnetic fields in your home: proximity to power lines or transformers; neutral currents on your own building's water pipes; wiring errors in some circuits; neutral currents on street water pipes that can circulate through your home; and wiring that employs the old knob-and-tube method. If you detect high magnetic fields in your home (anything consistently above 3 milligauss) and do not live near a high-tension line or transformer, the source might be your domestic wiring and its grounding system.

Electricity enters the home through what is called a service drop and flows through two energized ("hot") wires; plus a third neutral wire is connected back to the power line. The service drop is typically attached to a corner of the house or garage, and often a meter is nearby, with a metal wire extending from it that is driven directly into the earth or "ground" below. (As just mentioned, a ground is intended to direct unusual currents into safe places and to stabilize the entire network.)

Electricity flows two ways through this system. When you use appliances or lights, current flows through the wires, coming directly from the distribution lines along the street. As normal 60-hertz current reverses, or alternates, 120 times a second, return electricity flows back through the neutral line to the transformer on the outside electric pole. The neutral wire completes the circuit, and the system would not operate without it.

Each electrical outlet is also grounded to the system with small safety wires. It is this internal grounding equipment that carries extra or fault current back to its source efficiently enough to trip a breaker and stop the flow of current if need be. This prevents fires and electrical shock. GFI (ground fault interrupter) outlets are best designed for it and are required by code in new construction if an outlet is within several feet of a water supply, like the kitchen sink or a bathtub.

How internal wires are spaced can also be important. If the hot wires are separated from the neutral wire, and current is diverted away from the energized wires, high magnetic fields can result. The hot and neutral wires need to be kept close together so that cancellation of the fields can occur. Also, if the neutral current is allowed to take an alternative pathway back to the service entrance, the result is the same as with the separation of the conductors.

In theory, there is an orderly, safe electrical flow, but sometimes current ends up where it does not belong. Two kinds of current problems can elevate magnetic fields. One is internal current following along pipes or other metal conductors due to wiring errors. (This is always a code violation and should be corrected.) The other is a by-product of intentional grounding (not a code violation), which, depending on circumstances, can bleed off neutral currents that will then follow any metallic surface.

In other words, neutral currents can circulate in odd current loops that create high magnetic fields. Such current loops can follow any metallic pathway and have been found on cast-iron radiators, gas lines, water pipes, and aluminum window frames and in metal lathe and construction steel. (Plastic pipes, however, will not conduct electricity.)

There are several ways that neutral currents can be shunted

away from the neutral line that should be carrying them back to the pole transformer outside. One is the miswiring, misbonding, or incorrect grounding of an electrical subpanel. Another is when the neutral current is provided with an alternate path through some such mistake as a misplaced carpenter's nail. And a common way is for an electrician to "wire-nut" the neutral wires from different branch circuits together in the same junction box (against code).

High fields due to unusual wiring of subpanels primarily are found in newer homes or those to which additions have been added. A subpanel is located away from the main service box. Simple rewiring by a licensed electrician can fix this. If you get high readings near a subpanel, call your building inspector to check for violations. (Unfortunately, some inspectors don't realize they are looking at panel violations.)

Another way that unwanted current reaches a home is through common grounding practices for external metal water-supply systems. Most of these cases occur in urban or suburban neighborhoods whose water comes from a network of pipes running into apartment houses or single-family homes, rather than in houses with private wells. In such an environment, grounding problems can affect whole neighborhoods, because current will flow along your neighbor's pipes onto yours and continue all the way through the network. Sometimes you can detect such a situation if you get high magnetic-field readings on your pipes when your neighbors increase their electrical use, such as at dinnertime. With a proper gaussmeter, you can actually follow this field along the pipes to a neighbor's house.

Some homes with grounding problems like this report ambient magnetic fields of 20 milligauss and higher. One California study found magnetic fields as high as 120 milligauss in homes within a typical residential neighborhood. Clearly these are very high readings, and anyone living in such a high magnetic environment needs to investigate how to mitigate the exposure.

There is only one way to reduce ground-pipe problems. First

have the utility company clean and reconnect the neutral grounding conductor at the service drop to make sure there is a good
connection. Then have a plumber insert what is called a dielectric
union on the water-pipe shutoff at your property line, so that current from your neighbor's home cannot reach yours. A dielectric
union is a fitting filled with nonconductive material that will not
allow electricity to flow across it. This solution involves the hourly
fee of the plumber, as well as the cost of digging down to your
water-supply pipe. You should also investigate code regulations in
your area to make sure that this is permitted.

A similar solution that is allowed by code is to have the
plumber insert a length of plastic pipe (about one foot) into the
metal pipe about ten feet away from the house. This is also a good
solution if your soil is constantly damp, since dissolved salts in
the soil conduct electricity.

If you detect high background fields in your home and you
suspect it is from ground-plumbing problems, perhaps the best
approach is to hire an EMF consulting firm for diagnosis and
mitigation advice. Such a consultant will be able to instruct electricians and plumbers—who often do not understand the problem—on what to do. Increasingly, EMF mitigation firms have
electricians on staff, which simplifies things. Very simple alterations can sometimes bring dramatic changes in readings. You
can literally fix this problem within one day.

Other Wiring Sources of High Residential Fields

Wall wiring that is against code can also be responsible for
high residential magnetic fields. Such violations are so common
that many electricians and building inspectors are not even aware
of them. One way to identify such a problem is when your gaussmeter detects a higher-than-expected loop around one or two
walls of the same room. If you have not found high fields near
your service drop or water pipes and do not have an electrical

heating system installed in either a floor or a ceiling but still detect high fields within your living space, wall wiring is probably responsible.

Some electricians take the cheap and easy way of doing things, without understanding the later ramifications. One common source of high residential fields is the improper wiring of three-way light switches, the type that controls one light from two or more locations. Sometimes an electrician will separate the energized wires from the neutral wire, running them on separate walls. (In technical terms, the electrician has used a two-wire "traveler" when he or she was supposed to use a three-wire. It is a code violation.) This effectively creates an EMF loop that can cause significant elevation of magnetic fields in common living space. The only remedy is to rewire the area, making sure that the hot and neutral wires are close together.

One way of detecting wiring problems is when you get unexplained interference on your television set. Sometimes interference can be a by-product of backfed current loops. Call an electrician if you suspect this. Correctly wired three-way switches never create unusual fields.

Unfortunately, a building inspector will not be able to identify most wiring errors without opening every electrical box in the house, so a gaussmeter is the best tool to locate problem areas. Once a problem has been identified, a knowledgeable inspector may be able to confirm the violation and suggest a remedy.

Some legal wiring methods can also elevate magnetic fields. The old method of wiring (called knob-and-tube) is no longer allowed by code but still exists in a grandfathered status in older homes. It was used throughout the 1940s, and even into the 1960s by some electricians. Knob-and-tube configurations separate the "hot" wire from the neutral wire by a foot or more, running them on separate joists or rafters. (The knob is the porcelain connection of the wire to the rafter or joist; the tube is the porcelain tube through which the other wire passes through the rafter or joist.) It is the separation of the wires that creates the high magnetic fields.

If you live in a home built prior to the 1950s, you may have knob-and-tube wiring. It can usually be identified with a gaussmeter, or sometimes at the electrical panel box—the old conductors for knob-and-tube are dark and have extremely thick insulation. The only way to mitigate for elevated magnetic fields due to knob-and-tube configurations is to rewire the house, and the expense can be substantial. Rooms that are seldom used might be left alone, but that is an individual judgment call.

If you have eliminated all these possibilities for high magnetic fields in your home and still come up empty-handed for a cause, unfortunately dangerous wiring-code violations throughout the house may be at fault. This can be serious for a number of reasons, including the fire risk. In most cases, one or more wiring-connection errors are responsible and can be found and corrected.

People who live or work in multistory complexes and find extremely high magnetic fields should look to the location of the building's transformers and switching cabinets, as these can create tremendously high fields in adjacent apartments or offices. Sometimes wiring or rephasing corrections can help, as can a rearrangement of the living or working space. And passive shielding materials are available, such as low-carbon steel or mu metal. An EMF mitigation firm should be able to advise you on these.

High magnetic fields due to current problems or diversions in a home or work environment deserve investigation since they can often indicate fire hazard as well. It isn't something to be ignored.

If you have an electrician you like but who isn't especially knowledgeable about magnetic field problems, you might get him or her an excellent book about magnetic-field mitigation, written for electricians, although in language understandable to the nonprofessional. It is called *Tracing EMFs in Building Wiring and*

Grounding, by Karl Riley, and as of now is the only book of its kind. It can be ordered directly from Magnetic Sciences International, HCR-2, Box 850-295, Tucson, AZ 85735; phone 800-749-9873. You might give a copy to your local building inspector, too. Another useful reference for an electrician is an article that appeared in *Electrical Construction & Maintenance (EC&M) Magazine,* March 1993, Volume 92, No. 3, called "Magnetic Fields From Water Pipes," by Fred Hartwell. EC&M is an Intertec publication whose number is 913-967-1801 for reprints.

A Quick Note about Underground Wiring

Sometimes burying electric cables can reduce electromagnetic fields, but not for the reason that people think. The layer of earth is not what reduces the fields, but, rather, the fact that, in order to minimize digging, the cables are wrapped or brought close together, thereby canceling out some fields.

But burying cables can sometimes increase exposures, too, especially if you spend a lot of time near them. That is because underground domestic cable is typically buried just below the frost line in the northern states (about three feet down), and often less than that in milder climates, whereas overhead cable strung to the service drop must be at least ten feet above the ground.

If you live near buried electric cable, measure it with your gaussmeter, or ask the power company to do so.

Electrical Appliances

The average American home has approximately thirty electrical appliances and tools, many of which produce magnetic fields well above a few milligauss in the 60-hertz frequency. The ambient magnetic-field background in most American homes ranges from 0 milligauss to 4 milligauss away from appliances, with an average of about 0.38 milligauss. While most mitigation experts say that this all-pervasive electromagnetic environment is mostly influenced by wiring or grounding problems or the presence of power lines, appliances contribute their share when someone is near them.

The actual EMFs in any given room will vary according to the number and kind of sources, distance from the source, and how many of the sources are operating at the same time. The EPA measured many common appliances and reportedly was surprised at how high some of the readings were. The measurements were conducted on a number of products within each category, which the EPA ranked in a low, median, or high range, but it did not publish specific manufacturers' data. So we are still left to make our own measurements of our own appliances.

The following EPA tables show measurements made at various distances from the source. The dash (—) indicates that the magnetic field at that distance could not be distinguished from the background measurement taken before the appliance was turned on. The magnetic-field strength is given in milligauss (mG).

Some of these readings might give pause to people in certain occupations, such as carpenters, who use power tools regularly, cabinetmakers, who are surrounded by powerful machines not even tested by the EPA, and professional chefs, who work with large industrial mixers, processors, electric stoves and ovens, and refrigerators.

The strength of magnetic fields from vacuum cleaners was surprising. Central vacuum cleaners, however, with the motorized unit located away from the hose and wands, would have the highest fields near the motor, while the hose handle would be significantly lower. The same holds true for central air conditioning and heating systems. The highest fields would be near the main unit or condenser, not near the ducts from which the air flows. Also, many heating companies offer electronic air filters that zap dust particles from the return ducts to clean the air. Such an electrified element would add to the magnetic fields already

EPA TABLES

Distance from Source (Strength in mG)		6"	1'	2'	4'
BATHROOM					
Hair dryers	low	1	—	—	—
	median	300	—	—	—
	high	700	70	10	1
Electric shavers	low	4	—	—	—
	median	100	20	—	—
	high	600	100	10	1
KITCHEN					
Blenders	low	30	5	—	—
	median	70	10	2	—
	high	100	20	3	—
Can openers	low	500	40	3	—
	median	600	150	20	2
	high	1500	300	30	4
Coffee makers	low	4	—	—	—
	median	7	—	—	—
	high	10	1	—	—
Crock-Pots	low	3	—	—	—
	median	6	1	—	—
	high	9	1	—	—
Dishwashers	low	10	6	2	—
	median	20	10	4	—
	high	100	30	7	1

EPA TABLES
(continued)

Distance from Source (Strength in mG)		6"	1'	2'	4'
KITCHEN *(continued)*					
Food processors	low	20	5	—	—
	median	30	6	2	—
	high	130	20	3	—
Garbage disposals	low	60	8	1	—
	median	80	10	2	—
	high	100	20	3	—
Microwave ovens	low	100	1	1	—
	median	200	40	10	2
	high	300	200	30	20
Mixers	low	30	5	—	—
	median	100	10	—	—
	high	600	100	—	—
Electric ovens	low	4	1	—	—
	median	9	4	—	—
	high	20	5	1	—
Electric ranges	low	20	—	—	—
	median	30	8	2	—
	high	200	30	9	6
Refrigerators	low	—	—	—	—
	median	2	2	1	—
	high	40	20	10	10

EPA TABLES

(continued)

Distance from Source (Strength in mG)		6"	1'	2'	4'
KITCHEN *(continued)*					
Toasters	low	5	—	—	—
	median	10	3	—	—
	high	20	7	—	—
LAUNDRY/UTILITY ROOM					
Electric clothes	low	2	—	—	—
dryers	median	3	2	—	—
	high	10	3	—	—
Washing machines	low	4	1	—	—
	median	20	7	1	—
	high	100	30	6	—
Irons	low	6	1	—	—
	median	8	1	—	—
	high	20	3	—	—
Portable heaters	low	5	1	—	—
	median	100	20	4	—
	high	150	40	8	1
Vacuum cleaners	low	100	20	4	—
	median	300	60	10	1
	high	700	200	50	10

EPA TABLES

(continued)

Distance from Source (Strength in mG)		6"	1'	2'	4'
LIVING/FAMILY ROOM					
Ceiling fans	low		—	—	—
	median		3	—	—
	high		50	6	1
Window air conditioners	low		—	—	—
	median		3	1	—
	high		20	6	4
Tuners/Tape players (including VCRs)	low	—	—	—	—
	median	1	—	—	—
	high	3	1	—	—
Color TVs	low		—	—	—
	median		7	2	—
	high		20	8	4
Black-and-white TVs	low		1	—	—
	median		3	—	—
	high		10	2	1
BEDROOM					
Digital clocks	low		—	—	—
	median		1	—	—
	high		8	2	1
Analog clocks (conventional face)	low		1	—	—
	median		15	2	—
	high		30	5	3

EPA TABLES
(continued)

Distance from Source (Strength in mG)		6"	1'	2'	4'
BEDROOM *(continued)*					
Baby monitors	low	4	—	—	—
	median	6	1	—	—
	high	15	2	—	—

Older conventional electric blankets were found by the EPA to be close to 40 mG at average peak, 2 inches away from the blanket. New models, called positive temperature coefficient (PTC) low-magnetic-field blankets, were much lower, at around 3 mG, but many would say that even that is still too high for such a long exposure.

TOOLS/WORKSHOPS					
Battery chargers	low	3	2	—	—
	median	30	3	—	—
	high	50	4	—	—
Electric screwdrivers	low	—	—	—	—
(while charging)	median	—	—	—	—
	high	—	—	—	—
Drills	low	100	20	3	—
	median	150	30	4	—
	high	200	40	6	—
Power saws	low	50	9	1	—
	median	200	40	5	—
	high	1000	300	40	4

created by the furnace itself. There are passive electrostatic filters that can be retrofitted onto furnaces at far less cost. They are reusable and washable and emit no EMFs of their own.

Another point that must be re-emphasized: magnetic fields penetrate walls, so the reading for an appliance must include an accurate assessment for a bed or crib or chair against the wall in the next room. The same principle applies to the floor over a furnace, hot-water heater, or electrical junction box. Don't forget to measure all floors for electromagnetic fields.

Stoves and Ovens

Several new EMF/RF and magnetic technologies are being offered today for cooking food: conventional coil stoves can give off strong EMFs within close proximity but much lower ones one to three feet away; radiant burners employ very thin coils underneath a glass ceramic surface that operate at a very high wattage, so EMFs can be higher than with standard coil models; halogen burners use tungsten halogen bulbs in the light frequencies under a glass ceramic surface that conduct heat directly to the pot; and induction smooth tops with coils use magnetic friction to heat pots directly without heating the glass surface. Since induction smooth tops use magnetic fields to heat, pots must be made of a magnetic material, such as iron or steel. Induction heating will create intense magnetic fields in close proximity to you and your family.

Controls are either continuously variable, which can produce a 180-hertz frequency, or notched, with specific settings that are usually in the 60-hertz range. Some models may use RF frequencies. Obviously, with these new models and options, there are several different frequencies being employed, sometimes within the same unit.

Regular coil models with notched controls are probably the least complicated from an EMF standpoint, and mitigation is easy with simple distance. For those who wish to avoid such appliances altogether, some recommend gas stoves. Newer gas stoves have electronic pilot lights that emit a negligible amount of EMFs; older models are strictly manual and emit none, unless they have analog clocks.

Ovens now use several of the same technologies as the new stoves. Convection models use an internal fan (with an additional motor creating additional EMFs) to circulate hot air around food in order to reduce cooking time, as against the conventional method of simply heating a chamber and letting the food cook slowly. It is a question of weighing the alternatives: one oven has higher fields for a shorter period of time; the other has lower fields for a longer period.

Some wall ovens also employ an internal fan (which means another blower motor), whereas freestanding stoves usually have a passive vent under one of the stove burners that allows heat and moisture to escape without an additional motor.

Self-cleaning ovens employ tremendous wattage to reduce cooking debris to ash, and during the self-clean cycles they emit very high EMFs. Plan your cleaning for a time when you do not need to be in the kitchen.

Microwave Ovens

Microwave ovens, which seem to be everywhere these days, are quite different from regular ovens. They function by heating the food itself from the inside out (every other kind of heating is from the outside in). The technology employs high-frequency radio waves in the microwave band starting at around 500 megahertz and extending up to the infrared frequencies.

The way it works is this: A magnetron in the back of the oven produces a beam of microwaves that strikes a spinning fan, which then reflects and diffuses the waves onto the food from all directions within the oven. The waves have a deep penetration ability, pass through food containers, and strike water molecules within the food. The microwaves force the water molecules to align, then rapidly reverse that alignment. It is the rapid, repeated twisting and reversing at the molecular level that produces heat.

There are several things to be concerned about with this technology, one of which is the microwaves themselves. The early ovens (those manufactured before 1973) were found to leak microwaves so badly that *Consumer Reports* recommended that people not use them. Since then, the ovens have become safer, but it is important to understand that they all leak somewhat. The FDA, which regulates the ovens, allows a leakage of 1 mW/cm^2 for ovens still in the store. Once you get one home, maintenance is strictly up to you. It is important to have a reputable repair person check your oven with a microwave meter annually. There is great variation in the accuracy and calibration of meters available to the public, and no accuracy standards for them exist as yet. Professionals often have better measuring equipment, although not always. If you find your microwave oven is leaking, replacing the gasket around the door will often do the trick. Do not use the oven until repairs have been made. No safe levels for microwave exposures have been determined, but one study by Dr. Authur W. Guy indicated that a level of less than 0.5 mW/cm^2 is probably required.

Another source of radiation from microwave ovens is the strong EMFs they emit in the 60-hertz range. They come from the transformer/magnetron motor, which draws a substantial amount of current where it plugs into the wall outlet. Measure this with your gaussmeter.

It is important to keep children away from a microwave oven while it is in use. The placement of the microwave oven is also important—many people tuck it under a countertop right near their food-preparation area, which means high exposures to different frequencies when the oven is on. If you must have a microwave oven, it is recommended that you position it away from your work areas and the general traffic flow. When you use it, hit the "start" button and then walk away.

Many combination stove models now come with a microwave oven stacked on top of a regular oven or the cooktop. It is probably better to have a freestanding microwave oven, so that you can keep it at a distance from the other appliances. Such combination models will limit your design abilities in this regard.

Very little research has been done on the actual quality of food cooked in microwave ovens, other than determining that salmonella bacteria may survive in some cold spots that do not fully cook. But some research indicates damage to microwaved food at the molecular level that can theoretically be passed on to those who consume it.

A few European researchers have focused on human blood samples in test subjects who ate microwave-prepared food, and their findings are disturbing. There was one study with a small number of test subjects, but its findings are not widely accepted, perhaps because few people know of it and it has not been replicated. Researchers Bernard H. Blanc, of the Swiss Institute of Technology and University Institute for Biochemistry, and Hans U. Hertel, of Environmental-Biological Research and Consultation, took blood samples from three groups of volunteers over a two-month period. Their blood was found to have significant alterations in several ways.

Eight test subjects who were all on a macrobiotic diet received vegetables and raw milk that was either uncooked, cooked by conventional heating methods, or prepared in a microwave oven. Blood samples were taken immediately before and at defined intervals after food intake. All the food that was heated, defrosted, or cooked in a microwave oven caused significant changes in blood chemistry. The changes included a decrease in all hemoglobin values (red blood cells, which carry oxygen to tissue and remove carbon dioxide), as well as increases in hematocrit (the percentage of red blood cells in the entire blood count), leukocytes (white blood cells that are instrumental in fighting infection), and cholesterol values of both the HDL ("good") and LDL ("bad") types. Also, lymphocytes (a disease-fighting form of leukocyte manufactured in the lymph nodes) showed a distinct short-term decrease after microwaved food was eaten.

In addition, the luminescence of certain bacteria in the microwaved food was heightened, leading the researchers to conclude that microwave energy may be inductively passed on to the consumers of the food. The alterations in blood chemistry indicated to the researchers the kinds of early pathogenic processes similar to those that occur in the beginning stages of some cancers. And they noted that similar blood changes often point to states of stress. They also observed changes in the milk itself in the microwaved samples. Acidity increased, protein molecules were damaged, fat cells enlarged, and folic acid decreased, among other things. Folic acid is a member of the vitamin-B family that has been found to decrease the incidence of spina bifida (a birth defect commonly called open spine syndrome) in the offspring of women who take it in sufficient quantities before pregnancy begins.

What this Swiss study means, no one knows. Although the test group and the test parameters were highly controlled, it was nevertheless a very small sampling. A larger study is in order.

In America, researchers at the Stanford University School of Medicine in Stanford, California, have noted that a mother's breast milk, when reheated in a microwave oven, loses some of its infection-fighting properties. Microwaving breast milk was found to break down antibodies and proteins that normally inhibit bacterial growth and help a newborn fight infection. Microwaving breast milk appears to compromise its immune-fighting abilities. (The practice of collecting and freezing breast milk, then reheating it in a microwave oven, has become standard in most neonatal care units today.)

In another experiment, the same Stanford researchers found that milk heated in a microwave oven and then dosed with *E. coli* bacteria grew the bacteria eighteen times more than nonmicrowaved milk did. Milk heated to a low temperature was still found to grow the bacteria five times faster.

Personal Appliances: Hair Dryers, Electric Razors, Electric Blankets

Many of the small personal appliances we have come to rely on can generate high EMFs. Hair dryers need a high current to produce heat and so can generate substantial magnetic fields around your head. A 1,200-watt blow dryer can produce a 50-milligauss field at six inches, and higher the closer it is to your scalp. Even at eighteen inches from the front of the blower, 10 milligauss is a common reading.

The only thing in the dryer's favor (aside from convenience) is that its use tends to be of short duration. Many people now say that hair dryers should not be used on children at all. Manufacturers should be pressed to produce safer models that are shielded inside to decrease the fields; one low-EMF model is said to be available. Another solution would be to design a consumer model similar to the dryers found in some hotels, which have a hand-held flexible blower separate from its wall-mounted motor. (The magnetic fields are near the motor.)

Hairdressers need to be especially careful about the way they use blow dryers. Often a hairdresser will cradle a working dryer against the torso, producing high exposures off and on all day in the vicinity of the breasts and reproductive organs. Hairdressers are also exposed to a range of toxic chemicals and fumes. Given what is suspected about EMFs as co-teratogens with such chemicals, additional EMF exposures should be mitigated. Rest any working blow dryer on a safe surface, not against the body, and try to work in a well-ventilated area.

Electric razors can also be a source of short, intense EMFs directly against the face. A 1992 study conducted by Dr. Richard Lovely of the Battelle Seattle Research Center found an increase in acute non-lymphocytic leukemia in men who used electric razors. Dr. Robert Becker recommends that men with facial moles not use electric razors, because in theory these moles can become melanomas.

Electric blankets, as we have already said, can create high magnetic fields (upward of 40 milligauss) directly on the body over a number of hours. Such an exposure may disrupt the normal nighttime melatonin production in the brain, but daytime use will do the same. There is a link between breast and other forms of cancer and melatonin suppression, and melatonin is also known to be suppressed in those with Alzheimer's. (See Chapter 10.) There is now ample evidence that electric blankets are detrimental, despite the availability of newer models with reduced fields. Some say that anyone who uses an electric blanket is foolish. Others say it is a question of risk/benefit. For example, an elderly person in frail health might be in more danger from extremely chilly nights than from EMF exposure. Pregnant women or those trying to conceive should not sleep under a working electric blanket, or use an electric heating pad. Prudent avoidance entails heating the bed before you get into it and then not only turning off the control but unplugging the blanket.

With any personal appliance, dose rates also come into play. For instance, hair dryers and electric razors are used for relatively short periods of time, whereas electric blankets are in use for six to twelve hours. Massage chairs, electric massage devices, and electric heating pads are also used for extended periods of time. Clearly the exposures are different under these circumstances. All exposures now appear to be cumulative. There is also increasing evidence that the short, intense exposures from such appliances may trigger important physiological parameters in humans. Some estimates hold that the intense fields of electric razors, for instance, are equivalent to a 24-hour, 2-milligauss exposure.

Fluorescent and Incandescent Lights and Light-Dimming Switches

Various kinds of lighting have different electromagnetic properties. Millions of people live or work under fluorescent lights, which create much higher magnetic fields than does incandescent lighting. The newer globe or U-shaped fluorescent bulbs made to screw into regular lamp sockets are being touted as environmentally better, since they use less power. While it is true that they are more economical, it is also true that such bulbs will bring strong magnetic fields much closer to the body—right next to a person's head, in the case of a reading lamp.

Incandescent bulbs consist of a tungsten wire filament that is wrapped in a tight coil. Electricity passing through the coil heats it literally white-hot, with the temperature reaching about 4,500 degrees Fahrenheit (2,500 degrees centigrade). (Tungsten is used because it has a very high melting point.) The globe of the bulb contains an inert gas, like argon, to prevent the heated metal from combining with oxygen in the air, which would cause the filament to burn out. Incandescent bulbs are a fairly straightforward affair. Unless they are connected to light dimmers, they function in the 60-hertz frequency.

Fluorescent lights are another technology altogether. Fluorescent fixtures contain a phosphor-coated glass tube that glows when electric current is passed through it. Electrodes at each end of the tube are heated by current and emit free electrons. The electrons strike atoms of mercury vapor contained in the tube, and this causes the atoms to emit ultraviolet light, which in turn energizes the phosphor atoms into emitting white light.

While the light from incandescent bulbs and fluorescents looks alike, it isn't. Incandescent bulbs emit a much broader band of the visible light spectrum (with concentrations in the yellow-orange range). Fluorescents focus on a narrow band in the blue-violet range and are often deficient in red-orange. (Some newer fluorescents do come close to the full daylight spectrum, however.)

The magnetic fields emitted by the two are different. Incandescent lamps function with a simple metal filament that glows when 120 volts of current passes through. Fluorescents, which lack filaments, need to have that 120 volts of common house current raised to several thousand volts by a transformer contained in the ballast of the light fixture. This ballast is formed of coils that create a significantly higher magnetic field than an incandescent light. A 10-watt fluorescent tube can produce a magnetic field at least twenty times stronger than an incandescent one. Fluorescents typically emit a 6-milligauss field at two inches, 2 milligauss at six inches, and 1 milligauss at three to five feet. A 60-watt incandescent light is around .3 milligauss at two inches, and 0.05 milligauss at six inches; the field becomes lost in the ambient background at one foot. The magnetic field from a fluorescent fixture depends on the length of the tube, and the exposure depends on the height at which it is mounted.

Whole banks of fluorescent lights, with several fixtures of 20 watts each—a common array in schools, stores, and office buildings—can produce fields in excess of 1 to 4 milligauss near the heads of those directly under them. Given the length of time often spent in such environments, exposures can be significant. In addition to magnetic fields, fluorescent lights give off a small amount of ultraviolet radiation, which is known to cause skin cancer. Although most of the ultraviolet light is supposed to be blocked by the glass of the bulb, some of it is also blocked by plastic covers or diffusers, but often these are removed in order to increase the lighting capacity. Such covers do not block magnetic fields, however.

An increased incidence of malignant melanoma has been noted in two studies, one in Canada, the other in Australia. Both found a doubling of melanomas in those exposed to fluorescent lights at work. (A subsequent Australian study did not confirm the findings of the first study. It concluded that an observed increase in some kinds of melanoma was attributable to chance.)

The FDA recently estimated that unshielded fluorescent lights could add from 8,000 to 19,000 new cases of skin cancer a year in indoor workers exposed to them over a lifetime. That is a small number in comparison to the cancers thought to be caused by exposure to sunlight, but the FDA was not thinking of magnetic fields when it came up with those estimates, only of ultraviolet light. As discussed in Chapter 10, different causal agents are suspected with melanoma and EMF bioeffects, and the magnetic fields may prove to be more detrimental than ultraviolet exposure.

Fluorescent lights also cause headaches, eyestrain, concentration problems, and nervousness in some people, perhaps because the alternating current causes the lights to flicker. They are suspected of causing epileptic seizures in those prone to them, and one researcher is convinced that the abnormal light output of fluorescents causes behavioral disturbances in children whose classrooms have such fixtures.

Men with receding hairlines may be at higher risk from overhead fluorescent lights. If you have fluorescents at home, try to convert to incandescent lighting or keep at least six feet away from fluorescents. If you work under them, be sure they are covered. Keep a distance of at least six feet between the top of your head and the fixture. If your children's schools use fluorescents, try to convince the local school board to switch to incandescent bulbs. At the very least, make sure the fluorescents are covered and are mounted a minimum of ten feet from the floor, preferably higher.

Some new fluorescent lights on the market use electronic ballasts rather than magnetic ones. There also are new ways of mounting them, such as running two banks of lights on each ballast or positioning the ballasts back to back to promote field cancellation. These will reduce the magnetic fields, but electronic ballasts create significant 180-hertz harmonics, and they also give off RF, which may affect us in other ways. Some people are developing flu-like reactions in the presence of the electronic models. There are also prototypes of microwave light bulbs that are more powerful than microwave ovens. It will be difficult, if not impossible, to shield them for safety. Microwave lights are not a good idea.

TVs and Computers

Today's home typically has several television sets, and family members spend many hours a day in front of them. We have TVs in the kitchen, the living room, and the bedrooms, allowing us to watch morning, noon, and night. The average viewing time is four to six hours a day in most households.

While many people understand that TVs give off a small amount of ionizing radiation in the X-ray band near the screen, few know that the entire set emits a broad band of nonionizing frequencies that span the 60-hertz power frequency, the radio frequencies in the megahertz bands, and the light frequencies. The larger the screen, the stronger the fields will be, unless it is a large projection-screen model, in which case the stronger fields will be near the projection unit.

TV sets receive a video signal from a television station or video recorder. The set scans separate horizontal lines in order to form a complete picture on the screen. Electron beams sweep constantly from left to right; the fly-back circuit or wheel returns the line-sweep back to the left side at the end of each line. Color screens contain tiny stripes of phosphors that light up in red, green, or blue hues and combine to form the array of colors you see. Some color sets have three electron beams that scan for each color; in black-and-white sets, there is only one electron beam.

People often think that most of the emissions are from the front of the screen, but in fact the entire unit gives off varying fields. The strongest emissions are near the fly-back circuit, at around 15.5 kilohertz, but the entire set contains circuitry that generates electromagnetic fields in all directions. Remember that magnetic fields penetrate just about everything, so be sure not to position the back of a set against a wall that has a bed, a crib, or a frequently used chair or work space on the other side. It is also important to figure out where the 1-milligauss level is in front of the set. The larger the screen, the farther out it will be. Use a gaussmeter to measure the distance, or an AM radio to estimate it (as described in Chapter 11). In homes with children, the line can be marked with a small piece of tape on the floor.

Children should be kept back from TV sets, and from computer monitors as well. They tend to sit much too close to the screen while playing interactive games. The FDA has even issued warnings to parents, recommending at least three feet between the child and the TV set. But not all sets are the same; EMFs will differ from model to model. Take your own measurements—and be sure to include all the sets in the house.

Computers with cathode-ray-tube monitors have a radiation pattern similar to that of televisions. The biggest difference is that most computer users sit within a foot or two of the screen, especially if the keyboard is attached. Various computer models are discussed in the next chapter, but the same recommendations made for TVs also apply to personal computers.

Phones, Remote Controls, and Wireless Technologies

There has been a veritable explosion of wireless devices within the last decade, and more are constantly being developed, even without a clear consumer demand. If the major telecommunications companies have their way, within the next ten years many appliances and devices that are now wired (telephones, TVs, computers, modems, faxes, and the like) will become obsolete. That is not good news, because the new wireless models will function in the radio-frequency/microwave bands, as do the many wireless devices we already have on the market.

Called personal radio-transmitting devices, they include CB (citizens band) radios, walkie-talkies, cordless and cellular phones, some security systems for homes and businesses, remote-controlled toys, and a host of remote-control devices like automatic garage-door openers, TV/VCR channel changers, and wireless audio speakers. Although the frequencies and the transmitting power differ between these devices, the method by which they accomplish their action at a distance is by generating an electromagnetic wave that can easily pass through matter, including the human body. (Do not point a remote-control device at someone and press the buttons; an RF signal will go right through the person's body at close range.)

Phones: Many people today are confused about the difference between a cordless phone, which needs to be near a domestic telephone line of some sort; a cellular phone, which does not; or a car phone, with an antenna mounted outside the vehicle. Aside from being wireless, how do they differ from old-fashioned phones, and from each other?

Traditional telephones are all wired and use low-voltage direct current. Information comes and goes via a copper-wire system that is connected to other systems throughout the country and most of the world. Originally, telephone cable was laid across the ocean floor and strung on poles in every neighborhood to connect the systems. Today, transcontinental and transoceanic connections are by satellite. The handset of the telephone contains a microphone in the mouthpiece and a tiny speaker in the earpiece, which generates a small magnetic field. A small vibrating diaphragm changes the resistance of certain granules in the microphone, which causes the current flowing through it to vary. Dialing a predesignated number will connect two phone units through a network of exchanges. The sound signals in the handset are electrical signals carried on copper wires. When sent to the larger network, they can be converted into other frequencies: into light signals in fiber-optic cable or into radio frequencies via radio or microwave links. This is the way a wired phone reaches a car phone, for instance.

Wireless phones are similar in technology, with one exception. Their sound signals are transmitted in the RF/MW frequencies, with the transmitter in hand-held units often very close to

272 / ELECTROMAGNETIC FIELDS

the user's head. In cordless phones that are linked to domestic lines through a small base unit, this is done at relatively low power (less than 1 watt, depending on the model). Cellular phones, which are not linked with any lines, need more power and function between ½ and 6 watts of output. Such devices transmit in the microwave frequencies and can emit more radiation than the FDA allows for microwave ovens. Cellular phones are currently exempt from FCC regulations.

Then there are the mobile systems. Land-mobile systems operate through three kinds of RF transmitters: base-station transmitters, which are antennas usually inaccessible to the public that transmit over a large area and therefore require a great deal of power; vehicle-mounted transmitters, which can also require more power, although far less than a base station; and hand-held units—walkie-talkies and cellular and cordless phones—which need less power but nevertheless transmit at high frequencies.

Walkie-talkies typically transmit to other walkie-talkies, and the transmitter and receiver are on the same unit. The illustration on the following page shows the typical microwave exposures from a walkie-talkie. Cordless phones transmit to a base unit near a phone line but nevertheless require a transmitter near the head; under some circumstances they may actually turn the phone wires into RF-generating fields. Cellular phones use a distant base unit and require more power to reach it. Newer cellular phones have been proposed that would use even more power and would use satellites as their base, but there is serious concern over the health ramifications of such products.

Car phones are a combination of technologies. Unlike cordless phones, many car phone units have a handset connected by wire to equipment that is powered by the car battery but use an RF antenna mounted outside the vehicle for transmission. This at least gets the transmitter a little distance away from the user. Although a vehicle-mounted antenna has considerably less transmitting power than a tower base unit, it still requires more transmission power than a cordless domestic unit. It's six of one, half a

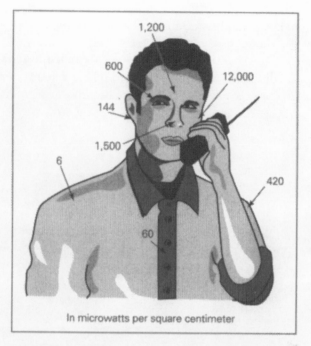

Typical Microwave Exposures from a Walkie-talkie
(EXPOSURE ESTIMATES ARE ADAPTED FROM *THE BODY ELECTRIC*,
BY ROBERT O. BECKER, M.D., AND GARY SELDEN, AND *ELECTROMAGNETISM
AND LIFE*, BY ROBERT O. BECKER, M.D., AND ANDREW A. MARINO, PH.D.)

dozen of the other. Cordless units are lower power but closer to
the user; vehicle phones are higher power but slightly farther
away.

Some vehicle models are wireless cellular units at the handset and use the outside antenna as a booster. These are probably
the worst exposures. It is possible to replace the cellular handset
with a wired one, and, with any vehicle model, it is best to position
the equipment in the trunk and use only a wired handset inside.
The antenna is best mounted on the roof but not directly above
the driver's head. Antennas mounted on the roof will radiate little
or no RF inside the car; those mounted on the trunk or side may

radiate inside. The illustration shows the exposures from a CB radio or car phone.

At least one manufacturer recommends that the user and anyone standing near a vehicle-mounted phone system maintain a distance of at least a few feet from the antenna during transmis-

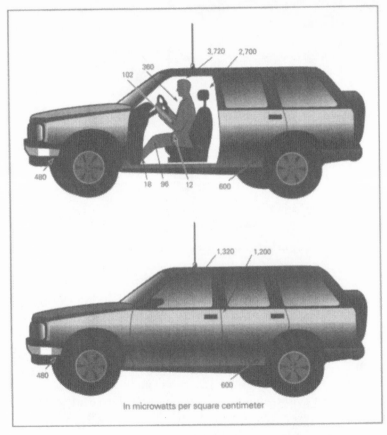

Typical Microwave Exposures from a CB Radio or Car Phone
(EXPOSURE ESTIMATES ARE ADAPTED FROM *THE BODY ELECTRIC*,
BY ROBERT O. BECKER, M.D., AND GARY SELDEN, AND *ELECTROMAGNETISM
AND LIFE,* BY ROBERT O. BECKER, M.D., AND ANDREW A. MARINO, PH.D.)

sion (anytime the user presses the "send" button). Most people do not realize that such devices affect those around them. In some instances, exposures can be greater to those nearby than to the actual user. Even domestic cordless phones can create a transmission range wide enough to produce a detectable signal as far as fifty feet away. There are reports of people with scanners picking up their neighbors' cordless-phone conversations, which indicates that the RF signals are creating exposures that far away. These are involuntary exposures. Those who live near busy highways are involuntarily being exposed to intermittent RF/MW radiation all day from car phones, as are the people in other vehicles. If you use a wireless phone of any type, it is not just you who is being exposed to their frequencies.

Cordless Phones and Specific Absorption Rates

The subject of specific absorption rates (SARs) is an important aspect of cellular phone use. SARs, which is shorthand for how much energy a particular area of the body absorbs, can differ greatly according to the area of the body being exposed and its particular tissue type.

As discussed in Chapter 3, the distribution of SARs inside the body is highly nonuniform, depending on such variables as the size, shape, and electrical properties within the body; the frequency of the wave, its intensity, distance, and polarization; and whether there is a reflecting surface nearby to direct waves where they do not belong.

The human anatomy absorbs radiation most efficiently between 70 and 87 megahertz in the FM band, but specific areas absorb radiation differently. The head and torso, for instance, can develop high local SARs at frequencies between 30 and 300 megahertz. The exposures that typically result from the higher frequencies used in portable transmitters like wireless phones can create SARs two hundred times higher in the head than in the rest of the body, according to a 1989 World Health Report. The body is resonant and can act as an antenna.

American cellular phones transmit at frequencies in the microwave range between 800 and 900 megahertz. (European systems operate at 450 megahertz.) Research has found biological effects at these frequencies. Swedish researchers found penetration of the blood-brain barrier, which protects brain tissue from injurious toxins, in lab animals exposed to frequencies near the cellular-phone range. Changes in calcium ions at the cell's surface have also been noted in ranges just below those of cellular phones, as well as at 450 megahertz, the frequency at which most American security radio phones function. And an increased incidence of leukemia has been noted in ham-radio operators, whose equipment transmits at around 145 megahertz.

Important research on the effects of RF/MW frequencies at the body's cellular level is also accumulating. Drs. Henry Lai and Narendra Singh, at the University of Washington in Seattle, recently found single-strand DNA breaks in the brain tissue of test animals after a single two-hour exposure to 2.45 gigahertz frequencies (in the microwave range) at power levels currently considered safe by the ANSI standards. Such DNA damage may be related to cancer initiation if an error in the repair process occurs. Also, Dr. Soma Sarkar of the Institute of Nuclear Medicine and Allied Sciences in New Delhi, India, recently found DNA rearrangements in the brain and testicular tissue of mice following microwave exposure at the same frequency and intensity as the Lai-Singh study. Replication of the Sarkar study, using the cellular phone frequencies, is planned at an American research facility soon, but replication of the Lai-Singh study, due to no fault of the researchers, appears paralyzed by typical EMF politics.

These are not the first researchers to report that microwaves can cause DNA breaks. In the mid-1980s, Drs. Jose-Luis Sagripanti and Mays Swicord at the FDA, working with Dr. Christopher Davis at the University of Maryland at College Park, reported that microwave radiation could cause both single- and double-strand breaks in DNA by acting synergistically with copper. This was an interesting observation with far-reaching implications for human health since the body is filled with traces of copper that are in fact necessary for good health. In addition, researchers at the University of Maryland School of Medicine found that microwaves can alter DNA in cell cultures after exposure to 2.45 gigahertz radiation modulated at 120 hertz. Drs. Elizabeth Balcer-Kubiczek and George Harrison reported that such irradiated cells became more cancerous if they were exposed to a tumor promoter afterward.

These DNA studies, if replicated and verified, may prove the mutagenic potential of microwaves hinted at in other work, as well as help overturn the ANSI standard. Their implications are critical to the cellular and cordless phone industry, as well as the other industries that profit from RF/MW devices or technologies—radio and TV broadcasters, for example. (For additional information on RF/MW research, see Chapters 13 and 15.)

Recent research has also focused on RF/MW absorption in the brains of cellular-phone users, but it has produced widely differing results, with variations of more than thirtyfold among some laboratories testing different phone models. Some test models used a head-shaped mannequin filled with a solution that simulated brain tissue to determine a worst-case scenario. Another researcher tried to take into account the shielding properties of the ear and the electrical properties of the skull bones in arriving at a "realistic" picture.

The latter approach was used by Dr. Om Gandhi, at the University of Utah in Salt Lake City. Gandhi's study was funded in part by McCaw Cellular Communications, Inc. and found, after tests of ten phones, that peak SARs and RF exposures were "well within national safety standards." Other researchers have reached the opposite conclusion. Dr. Niels Kuster, of the Swiss Federal Institute of Technology in Zurich, in worst-case scenarios, found SARs to be over thirty times higher than Gandhi's and concluded that today's phones are questionable with regard to safety limits. Dr. Quirino Balzano of Motorola (which manufactures cellular phones) found test results that coincided with Dr. Kuster's.

Critics of Dr. Gandhi's study questioned the validity of his findings on several grounds, including the fact that he did not press the cellular phone against the ear of the simulated human head—what person using a cellular phone does not press it to his or her ear? Also, the findings were announced in a press release rather than undergoing peer review. The press release was also misleading in a number of key areas, such as the statement that it was funded by the National Institutes of Health (NIH), which it was not. Both the FDA and the National Institute of Environmental Health Sciences (NIEHS), which is part of the NIH, have distanced themselves in the past from the kind of safety assurances that Gandhi espoused. The FDA, after Gandhi's press release, noted that it has had reservations for some time about the adequacy of the ANSI standard to which Gandhi was comparing the phones. Also, critics pointed out that Gandhi's RF exposures were "low" only when the cellular phones were tested under best-case conditions. After criticism of this kind from a number of sources, Gandhi reworked his original exposure estimates for the phones he had studied and found them to be too low by a factor of 1.08 to 2.47. He later informed the FCC that his original study contained a number of errors and that the peak SARs were higher. Prior to this, Gandhi had noted that antenna design can also be a factor in higher exposures: the smaller the antenna, the higher the SARs and the possibility of substantial coupling to the head.

Many studies have noted deleterious effects that were frequency dependent only and had nothing to do with power output. This, too, is of concern with radio devices. Currently, even the inadequate ANSI standard adopted by the FCC does not include low-powered devices (under 7 watts) like cellular and cordless phones. No federal standards for personal RF-transmission devices exist, although the FDA has them under its purview.

The German government now requires the testing of cellular phones under worst-case conditions to cover all the contingencies, but German standards are even more lenient than the ANSI standard. Other countries are developing their own test models and calculations for SARs. These include Australia, Austria, Belgium, Britain, Canada, France, Italy, and Japan. The science, however, is still evolving, and reliable results will take a long time to accumulate.

So what to do in the meantime? Some say not to use cellular phones or similar devices except in an emergency, such as being stranded on a highway. Others say, with regard to cordless domestic phones, to answer them if you are outdoors, but continue the conversation on a wired model. The FDA has issued an advisory warning that hand-held cellular phones should be used only when necessary and conversations kept brief.

There are important differences between wireless phone models, but no manufacturers were named in the studies just cited, probably because everyone was afraid of the liability. In general, worst-case models would be those in which the antenna is directly against the head. The shorter antennas on some newer models may also be more detrimental. Manufacturers may discontinue them in favor of longer antennas. Some manufacturers are experimenting with shielding the antennas. This may help, but such models will then require a greater power output, which could cancel the shielding benefit. Some semiportable models are available, though, that have the antenna and power supply in a separate tote bag, allowing the handset to emit no radio waves.

Prudent avoidance is to use such devices as little as possible, and not to let children use them at all except in an emergency. They are not like conventional phones, and we should not think of them that way. There is very little certainty about their safety, and as of this writing more than a dozen lawsuits have been filed involving cellular phones and brain tumors.

Cellular phone towers are discussed in the last chapter.

Chapter 13

WHAT'S AT WORK?

Most people spend over a third of their time at their place of employment, making the workplace a significant environment for possible EMF exposure. In fact, many occupations and work settings can expose employees to extremely high EMFs for long periods of time, whether or not the work is considered high risk.

Just getting to work can expose someone riding a train or subway to high EMFs from the trains themselves and from utility lines paralleling the track. Some trains can reach magnetic fields as high as 150 gauss on acceleration—a field that will cause paper clips to stand straight up. The new high-speed Maglev (magnetic levitation) trains operating in parts of Europe and in Japan use extremely high magnetic fields to lift the train and float it along the track. Magnetic fields similar to those used by Maglevs have been found in animal studies to decrease seratonin-N-acetyltransferase (NAT), an enzyme that controls melatonin production in the pineal gland.

Today's workplace contains occupational exposures that go far beyond what has traditionally been considered dangerous, like pollutants, chemicals, and ionizing radiation. Many white-collar jobs, once considered relatively free of dangerous substances, today are high-risk professions when it comes to EMF exposure. The typical office is high on the list.

Offices

Modern offices are awash with EMF-creating technologies that just a few years ago did not exist. And many of these new technologies operate in the higher infrared/laser/light frequencies. Cathode-ray tubes in computers and TVs can throw off ionizing particles in the X-ray bands; many offices are lighted by fluorescent fixtures that create high magnetic fields and ultraviolet radiation. There is a veritable symphony of electromagnetic noise from electronic equipment alone. In addition, offices located in high-rise structures can be affected by their proximity to electrical transformers, cables, and switching apparatus that many employees remain unaware of until an illness cluster appears or someone decides to take background measurements.

The trick with office equipment is to know what you are working with and to understand how proper positioning of the equipment can help mitigate the risk. VDTs have gotten much of the press coverage over the years, so most people think that their greatest EMF exposure is from computers. In fact, VDTs rank fourth as an EMF source, behind fluorescent lights, copiers, and building wiring. Many other machines contribute high amounts of EMFs, including some old electric typewriters. Sometimes just rearranging the work space can greatly reduce employee exposure, especially in a setting filled with rows of computers, like a newsroom or a data-entry operation.

Computers. Computers today come in many varieties and models, with great variations in EMF emissions. (Sometimes different machines of the same model have large variations.) The good news is that most manufacturers, in the absence of any U.S. standards, now adhere to the strict Swedish MPRII guidelines for emissions (less than 2 milligauss at 30 centimeters of extremely low-frequency EMFs from the screen), so it isn't difficult to find a new low-EMF monitor. But that may not be so if you are working with a CRT model that was manufactured prior to about 1989. The situation serves to show that manufacturers can create safer products if given the parameters and the motivation. They are to be commended for their action, even though it came after a frustrating period of stubborn public denial. Some manufacturers, however, sell their lowest-EMF models only in Europe.

In looking for new equipment, check to see if it complies with the TCO recommendations of the Swedish Confederation of Professional Employees—which represents over a million Swedish white-collar workers and has been a leader in the development of strict limits on VDT emissions. Some American models meet these standards, and say so in their support literature. The TCO standards are even stricter than those of MPRII; they include energy-efficiency requirements as well.

Not all older computers create high fields, but many of them that do are still in operation, and no manufacturer has recalled them. Often such items are donated to schools and churches, where they are used by children. All older models should be tested.

There are essentially three different types of computer screens. The cathode-ray tube is almost identical to that on a TV in that it uses a broad-band array of frequencies and a significant amount of power. The plasma screen relies on a radioactive gas pressed in between two panels, which lights up when bombarded by electrons. Plasma screens use a narrower range of frequencies but a significant amount of power. (They were originally devised by the U.S. military for pilots who needed high-resolution screens against the bright background of natural sunlight.) The liquid crystal display (LCD) and its various cousins, like the backlit VGA, use liquid materials that twist at different angles when polarized light and weak electrical current passes through them.

LCDs require much less power and use a still narrower range of frequencies.

Of all the screen types, the LCD is considered the lowest in EMF emissions (unless the screen is backlit by fluorescent tubes, in which case it may exceed the Swedish standard). This is true whether the LCD is color or monochrome; color displays in CRTs are higher in EMFs than monochrome models are.

The screen is not the only source of EMFs. Models with a lot of information-storage capacity in hard drives (30 megabytes and up) will give off higher EMFs. LCDs with a high storage capacity are found in some laptop models, which rest literally on the user's lap at genital level. Also, many of the newer programs offering windows and other computing capacities require a large hard-drive space. The prudent-avoidance advice is not to purchase more megabyte space than you really need; you can always up-grade a system later. Do not purchase a laptop with 120 mega-bytes of hard drive and 4 megabytes of RAM when all you really need is 20 megabytes of hard drive and 1 megabyte of RAM. The need is determined by the programs you want to use, and usually the space requirements are specified on the package.

This obviously applies to systems that you purchase yourself. But most employees work on machines provided by their employ-ers and do not have much choice. That is when taking EMF mea-surements becomes important (especially around mainframes), along with various machine-positioning strategies. Many employ-ers today are willing to reconfigure the work space to accommo-date the concerns of their employees. Others, unfortunately, are not, and unions are having to force the issue.

Many new computer systems are offering what is called a power-down feature—after a specified period of time in which a computer is turned on but not used, the power is automatically reduced. Although the feature was intended as an energy-saving device, it also reduces EMFs. If you habitually leave your ma-chine on all the time, this feature could be helpful. It also extends the life of the machine. It is still best, however, to turn off the equipment whenever you are not using it.

Printers. Printers also come in a wide variety of models using different technologies. The daisy-wheel printer, much like those on automated typewriters, has fairly high fields at the motorized spot that moves the wheel, but these drop off to near zero just inches away. Dot-matrix printers use closely spaced dots, which merge to form letters or numbers created by internal pins driven by electromagnets. They produce high fields near the control area, but these, too, drop off quickly with distance, although not as much as those of the daisy wheel. Thermal printers, which essentially burn an image onto a page, require a lot of power and create significant fields all around them. Laser printers, which use light frequencies, internal mirrors, and a rotating drum to create dots—something like a dot-matrix printer and a photocopier combined—also use a lot of power and create significant fields all around. There is a large variation in EMFs between manufacturers' models. The best course is to measure any equipment you are thinking of purchasing.

Copiers. Copiers can give off much higher EMFs than computers and, depending on the model, can utilize several frequencies besides the 60-hertz power frequency. Copiers function through a controlled system of mirrors, moving drums, lights, lenses, heaters, carrier belts, and toner powder, as well as magnets, which remove electrical charges from the drum, and a transfer charger, which applies a negative charge to the paper so that it will attract toner.

In general, the more capabilities a machine has, the higher will be the EMFs. For instance, a professional console-model printer that can stack, collate, and make copies at high speed will give off much higher fields than will a simple home copier (which can, however, be a significant contributor to background EMFs). But how one uses the equipment also counts. With a huge professional copier, operators typically program what they want and walk away until the job is done; with home copiers, the user tends to stand right over the machine while making copy after copy. No matter what kind of copier you use, it is best to step back about three feet when it is in operation. Do not stand close and make copies with the protective top lifted. Nor should you use a copier like a cheap camera to make pictures of your face or other body parts. Children should not be allowed to make multiple copies.

Faxes/Modems/Phones. Wired telephones are low-DC-current devices that are modulated with the voice frequencies (between 15 hertz and 1,500 hertz). They also have a 30-volt ring impulse, and a very small magnetic field at the earpiece that vibrates at voice frequencies. Faxes (short for facsimiles) and modems use telephone lines to send information from place to place. They utilize various frequencies and convert them to voice-modulated frequencies that can be sent over a phone line. Faxes in addition contain technology similar to a copier's and therefore give off higher EMFs within one foot than a wired phone does. Again, there is a wide variation in manufacturers' models, and the best thing to do is to measure your equipment. One day soon, telecommunications companies envision most of these items as being wireless, which means they will contain RF/MW transmitters. Because of serious concerns about the safety of this technology, it is recommended that people stay with wired models until more is known or strict national standards are in place. Also, keep in mind that you are not the only one affected by wireless items; so is everyone near you.

How to Arrange an Office

Before you commence to rearrange any work space, first find out what is in adjacent offices and storage areas. Sometimes the building's blueprints will help, and often the building supervisor will have this information. If you are adjacent to electrical apparatus, an elevator shaft, or main electrical conduits running under the floor or in the ceiling, professional measurements should be taken. Several clients in high-rise buildings have moved out of, and others have refused to lease space in, buildings with high EMFs. It is becoming an important health and environment issue with employees.

If you discover that your office abuts other offices with EMF equipment, try to move your desk, computer, and other equipment at least three feet away. Place bookcases or file cabinets there instead. Three feet away from operating equipment is the minimum distance that many now recommend, and sometimes even that is not enough.

Try to arrange your equipment in a row, rather than in a horseshoe configuration in which you sit in the middle, right where the fields from several machines may converge. Although the fields from office machines act individually, if the machines are too close together an augmentation effect may sometimes occur in a particular spot. There is not just one level of exposure; most machines give off rings of varying intensity that diminish (attenuate) with distance. What you are trying to effect is to work most of the time in a field of no more than 2 milligauss—preferably 1 milligauss.

Workstations should ideally have computers with detachable monitors that can be placed at least three feet away from the user (unless it is an LCD display). Disk drives should also be kept away from users, not right next to someone's legs. A printer should be on a straight line with the computer, not behind someone's chair, unless it is at least eight to twelve feet away. The same goes for copiers. Faxes, modems, and any other electronic gear should also be three feet away. The only thing that should be within arm's length is the computer keyboard.

Be aware of other people's equipment as well. Often, maximum exposures are generated from one side of a CRT computer, depending on how the fly-back transformer is arranged. Take measurements to determine where the lower 1-milligauss level is. In a setting with multiple machines arranged in long rows, try to position the monitors back to back and raised above head level.

If a woman worker is pregnant or trying to conceive, many women's health organizations recommend that she request a job reassignment until after delivery. At the very least, she should limit her computer use to no more than twenty hours a week.

Another mitigation recommendation is to turn off equipment when it is not in use, or purchase equipment with the power-down feature. Do not just leave the computer running all day whether you are using it or not. Also, stand away from equipment when it is operating. Just stepping back a few feet from a copy machine after pressing the button will help. And try to avoid a high cumulative exposure throughout the day.

Some companies offer antiglare screens that they claim shield against EMFs. But most such screens shield only for electric fields, not for magnetic ones, which can penetrate almost anything and are the component that is the most worrisome. A few manufacturers now offer metallic attachments that absorb EMFs, and these may be more effective. A few companies, such as Environmental Services and Products in Boulder, Colorado, make simple shielding cylinders that are inserted around the neck of a CRT inside the case. Another absorbing device, which attaches to the outside of the computer, is made by NoRad of Carson, California. It is a sort of metal headband that attaches to the sides of the computer and is said to absorb magnetic fields to below MPRII standards.* Other companies make cases that fit around various computer models, but these can be expensive. Sometimes it is more economical just to purchase a lower-EMF model.

---

* Environmental Services' product costs about $99 plus installation; phone 303-449-6780. For information about the NoRad device, phone 800-262-3260.

Some people claim that placing a large piece of natural quartz crystal on top of a computer will reduce EMFs, but there is no reliable information about the effectiveness of this, and no plausible theory that would account for a crystal's absorption of EMFs. Crystals are known to resonate at certain frequencies—like the crystals in old radio tubes—but that is a different phenomenon than absorption.

Low-EMF Architecture

Architects are paying more attention to low-EMF building design, partly because it is easier and less expensive in the design stage but difficult and costly after the building has gone up. In addition, some of the most prestigious professional addresses have been found to have extremely high EMFs, often unbeknownst to the landlords until it is discovered by existing or prospective tenants. For instance, according to *Microwave News*, a professional newsletter that reports on various issues pertaining to nonionizing radiation, Marine Midland Bank found high magnetic fields of up to 300 milligauss in some second floor offices at their 250 Park Avenue address in Manhattan. The bank filed suit against its landlord but dropped the suit when the landlord agreed to spend $1 million in mitigation and shielding costs. Other corporate clients, according to *Microwave News*, have also shielded office space. American Express employees reportedly noticed computer-screen jitter—a sure tipoff to high EMFs and electronic interference—at their corporate headquarters in lower Manhattan. EMFs as high as 105 milligauss were eventually found in some lobby offices, which were subsequently shielded. (Invariably, such high fields in a building are from electric-service equipment, the location of power cables, or electric switching gear.) Some corporate tenants have chosen to change addresses altogether because of EMF concerns.

Sometimes office exposures can be in the thousands of milli-gauss in some areas, and these are only for the 60-hertz fre-quency. RF frequencies on or within buildings can also be high. Cellular-phone and paging services are mounting their antennas on window ledges and sometimes are installing whole transmit-ters within elevator shafts. Anyone close to these will have a truly high intermittent exposure to several frequencies.

The World Bank, however, in designing its new corporate headquarters in Washington, D.C., built EMF mitigation for the 60-hertz frequency right into the design, making this the first major office complex to be specifically designed with EMF miti-gation in mind. The architects developed a number of prudent-avoidance features to limit exposure to magnetic fields from wiring configurations and transformers, as well as from such elec-trical equipment as computers, printers, copiers, fluorescent lights, and the like.

Their architectural designs placed the electrical-service equipment and office machines that generate the highest fields away from areas where employees spend the most time. They placed the main electrical switching equipment in a fifth sub-basement used for parking, rather than at street level, where it can generate magnetic fields in the hundreds of milligauss in ad-jacent areas. On floors where additional transformers are needed, they are being placed near elevator lobbies, in areas that will be used only intermittently, thus forming a buffer zone against the busy office space where people spend considerably more time.

The designers are also evaluating for EMFs all newly pur-chased office equipment, and are positioning desks away from fan-coil units used for ventilation. They have specified electronic ballasts, rather than magnetic ballasts, for fluorescent lights and are running two banks of lights on each ballast, thereby reducing the overall number of ballasts needed. In addition, they have run fixtures back to back so that the fields cancel each other out. Not only have these design features cost less to install throughout the huge city-block-size complex, but they are more energy efficient and emit much lower EMFs. (Electronic ballasts in light fixtures can emit significant 180-hertz harmonics, however.)

Many of those in the forefront of the EMF issue call the World Bank building, which is scheduled for completion in 1996, the office of the future. Architectural firms have also begun to send representatives to EMF conferences to learn more about mitigation near high-tension lines.

Remedying high-EMF problems in an existing building is not so easy or inexpensive. The cost of magnetic shielding materials made of special mu metal (iron alloys) and low-carbon steel can be astronomical, but a few landlords would rather mitigate than face empty corporate space. Most of these materials were originally designed to protect electronic equipment from electromagnetic interference, but they are increasingly being used to shield whole areas, because of concerns about employee health. Reducing the magnetic fields up to 10 milligauss will alleviate VDT interference, although this is still too high a limit for long human exposure. A limit of 5 milligauss will substantially increase the cost of mitigation, but a reduction to between 1 and 2 milligauss is more complex, since most office machines create fields higher than that within a one-to-three-foot radius. That is when low-EMF equipment and positioning help.

If you discover that your office or workplace is high in EMFs, a professional consultant will be able to diagnose the sources and make recommendations, which may entail anything from correcting a wiring error, to repositioning, to interior redesign, to shielding. The bill can range from a mere consultant's fee up into hundreds of thousands of dollars for shielding around electrical switching gear. But any company that takes its employees' health seriously will not expect them to work in abnormally high fields. And numerous lawsuits are making their way through the courts involving breast-cancer clusters, brain cancers, lymphomas, and leukemias that workers attribute to high EMF exposures in their offices.

Electrical Professions

A number of professions subject their workers to much higher EMFs than the average. Among these workers are electricians, electrical and electronic engineers, electronic-equipment assemblers, power, telephone, and cable linespeople, certain other utility-company employees (such as substation workers), switching-station operators, train and subway operators, motion-picture projectionists, welders, aluminum processors, certain workers in defense industries, and anyone who works with electrical equipment designed for the professional—kitchen workers using large mixers, processors, or multiple electric stoves or ovens; carpenters, cabinetmakers, and floor refinishers using electrical saws, drills, sanders, and so on. (Some of these occupations have cross-exposures to chemicals as well.)

There are wide differences in exposure within the job categories. Past studies often correlated mortality figures with job titles, but these may have provided only sketchy data. For example, electricians who work primarily on new construction typically do not work with energized wire; only a certain percentage of their work will involve higher EMF exposures. Those who hang telephone or TV cable also work with unenergized wire, but linespeople typically string cable along poles that already have energized lines, so their exposures will be greater. Cable splicers, who always work around energized wires, have the highest exposures. More recent studies, in an attempt at greater accuracy, are taking actual measurements of occupational exposure, rather than going by job title alone.

Numerous studies have found adverse health effects among such workers. Some studies have included adverse reactions in the spouses and children of workers, which indicate that the damage can be genetic and can be passed on to subsequent generations. Increased risks have been found for lymphomas, melanomas, breast cancer and other cancers of glandular tissue, and central-nervous-system cancers (of the brain) in particular.

Here are the results of some of the studies:

- Milham found in 1982 that electricians, power-station operators, and those who worked in aluminum processing (where substantial magnetic fields are created by metal-extraction techniques utilizing huge electromagnets) had a high increase in leukemia deaths.
- Wright found in 1982 a higher risk of acute myeloid leukemia in utility and telephone linespeople, as well as a general increase in leukemia in occupationally exposed workers in Los Angeles.
- McDowall found in 1983 an excess risk of leukemia in electricians, electrical and electronics engineers, and telegraph operators in England.
- Howe and Lindsay found in 1983 that Canadian transportation and communications workers with high EMF exposures were at a higher risk of developing twelve forms of cancer and of dying from leukemia. In addition, linesmen had more cases of leukemia and cancers of the stomach and intestine. Those in appliance manufacturing had a higher risk of dying from leukemia and lymphoma.
- Nordstrom found in 1983 abnormal development in fetuses whose fathers worked in high-voltage electrical substations.
- Nordenson reported in 1984 that workers in a Swedish 400-kilovolt substation, where they were regularly exposed to short, high magnetic fields and surges that caused spark discharges (called transients), had an increased rate of chromosomal breaks in the lymphocytes in their blood. (Transients may be an important factor in EMF exposure.)
- Lin found in 1985 that occupational exposures high in EMFs correlated with increased mortality from primary brain tumors (those that originate in the brain rather than metastasize from other areas). When grouped according to levels of exposure, electricians and electronic engineers were found to be at the highest risk. Those with the highest exposures had

twice the expected tumor incidence, as well as the highest mortality at a younger age. Lin also found in 1985 an increased mortality from cancers of the brain and liver, as well as from leukemia, in Taiwanese electric-power employees.

- Gilman found in 1985 an excess mortality from leukemia in miners exposed to EMFs from electrical distribution lines strung overhead in mine shafts.
- Pearce found in 1985 an excess of leukemia in New Zealand men with high EMF occupational exposures. Those at highest risk included electricians and radio/TV repair workers.
- Flodin found in 1986 an increased leukemia risk in Swedish electrical workers, including technicians, computer and telephone repair personnel, and welders.
- Stern found in 1986 an excess incidence of acute myelogenous leukemia in electrical workers in Sweden. Stern also found, in another study in 1986, a significantly increased mortality from leukemia in U.S. naval shipyard electricians and electrical welders.
- Savitz and Calle found in 1987 an excess incidence and excess mortality from leukemia in electrical engineers, electronics technicians, electrical-equipment assemblers, radio and telegraph operators, utility linespeople, welders, flame cutters, and aluminum processors.
- Szmigielski, Vagero, and Tornqvist reported in 1987, 1985, and 1986, respectively, an increase in skin-cancer deaths in EMF-related occupations, even in those who did not work outdoors in the sun.
- Thomas found in 1987 an increased risk of brain and other central-nervous-system cancers in electrical workers over the general population, particularly among those involved in the repair and manufacturing of electrical and electronic equipment. Brain-cancer incidence increased with the duration of employment exposure, and those with both RF and ELF exposures had double the risk for brain tumors.

- Coleman and Berel found in 1988 an 18 percent increase in leukemia in electrical workers.
- Juutilainen in 1988 reported a higher risk of leukemia in line-workers and cable joiners in Finland.
- Speers found in 1988 an increased risk of brain tumors in transportation and communications employees. Electric sub-station employees had a thirteen times greater incidence than expected.
- Linet found in 1988 a significantly greater risk of chronic lymphocytic leukemia in electrical lineworkers in Sweden.
- Mantanowski, in a review of an ongoing study, found in 1989 an increased risk of all cancers in New York telephone employees with EMF exposures. This included seven times as much leukemia as in control occupations, twice as many brain tumors in cable splicers (those with the highest exposures), and, for the first time, a significant increase in male breast cancer—an extremely rare form, whose appearance in the EMF-exposed population is important and has spawned further studies. Mantanowski's report also found dose ratios, meaning that the higher the exposure, the greater the risk.
- Mantanowski and Breysse found in 1989 that telephone cable splicers had a 70 percent increased risk of brain tumors.
- Johnson and Spitz in 1989 found that the children of fathers employed in EMF-exposed jobs had a 60 percent increased risk of brain and central-nervous-system tumors. Among electricians' children, the risk was three and a half times what could be expected. In earlier work, Johnson and Spitz found an association between childhood neuroblastoma risk and the father's EMF exposures.
- Savitz and Loomis found in 1989 that electrical workers died of malignant brain tumors at a 50 percent greater rate than men in other occupations. They also found an excess risk of leukemia deaths in electricians and electronics technicians.

- Reif in 1989 reported an increased risk of brain cancers in electrical engineers and electricians.
- Demers found in 1990 that breast cancer among male telephone lineworkers was increased sixfold over what would be expected.
- Mack, Preston-Martin, and Peters found in 1991 significantly increased numbers of primary brain tumors (astrocytomas in particular) in men occupationally exposed to EMFs, especially electricians and electrical engineers. A significant increase was found after ten years' exposure. They also found an increased risk of chronic myeloid leukemia in the welding profession.
- Garland found in 1991 an excess risk of leukemia in U.S. Navy electricians' mates.
- Loomis found in 1992 a doubling of breast cancer deaths among male electrical workers under the age of sixty-five. Loomis also found in 1993 that women in traditionally male-dominated electrical occupations had a 40 percent higher mortality from breast cancer than women who worked in nonelectrical jobs.
- Floderus reported in 1993 an elevated risk of leukemia and breast cancer in railway workers on electrified trains, although track and station employees did not appear to have an increased cancer risk. Engine drivers were found to have a threefold increased risk of chronic lymphocytic leukemia. In a re-examination of earlier work done by Floderus, it was also found that engine drivers had an eightfold increase in male breast cancer and that there was a fivefold increase for drivers and conductors. The study also found three times more pituitary-gland tumors than would be expected among drivers and conductors.
- Theriault, Miller, and Goldberg reported in 1994 that utility workers with a greater than average cumulative magnetic exposure were three times more likely to develop acute myeloid

leukemia than were less exposed workers. The study also showed that workers with the greatest exposures to magnetic fields had twelve times the expected rate of specific brain tumors (astrocytomas).

- London reported in 1994 an increase in all types of leukemia in electrical workers, which was dose and duration related. Workers who spent the most time in fields greater than 25 milligauss or who had worked in magnetic fields averaging more than 8.1 milligauss had a 2.3 times greater risk of developing chronic myeloid leukemia than the least exposed workers did. The greatest risk for all leukemias was among telephone lineworkers and splicers, at 3.2 times the expected rate.

- Sobel reported in 1994 an increase in Alzheimer's disease among workers with medium and high EMF exposures, in an analysis of data from Finland and America. He found that dressmakers and sewing-machine operators, with exposures near their heads of between 2.7 and 5.2 milligauss for home sewing machines and 2 to 11 milligauss for industrial machines, while at knee level the fields reached as high as 200 milligauss, were at 3 times higher risk for Alzheimer's in all sewers and 3.8 times higher risk in women.

- Dr. Zoreh Davanipour reported in 1986 a statistically significant increase in ALS (amyotrophic lateral sclerosis) in electrical professions where exposures exceeded 10 milligauss for twenty years.

Occupational studies often draw criticism from skeptics and cautionary interpretations even from their authors. Actual EMF exposures are difficult to assess: workers are sometimes exposed to a number of carcinogens, including chemicals and solvents, and researchers are not always able to understand what is biologically important in what they are measuring or how to interpret the outcomes. For instance, there is some indication that exposure to high-frequency transients—intense, short surges, pulses, or sparks of energy common in much electrical work—may be the important factor in adverse health reactions. The risk from transients may be higher than that from continuous 50- or 60-hertz fields alone. Animal studies and some human occupational studies seem to point in that direction. What is also worrisome about this small body of inquiry is that transients are similar in some ways to the pulsed electromagnetic fields often used in bone-regeneration techniques today and are typical of what we encounter when we use many small appliances. Transients can also affect melatonin production.

New studies come in every month; there are currently hundreds being conducted in various countries. The question is, how many studies are enough? We already have more studies finding adverse effects from high EMF exposures in the 60-hertz frequency than we do for smoking and various cancers. It is time we acted to protect certain occupations better than we do now, and certainly time to stop pretending that this body of work does not exist or requires endless verifications.

The Swedish Trade Union Confederation has already developed criteria for occupational exposure requesting that:

- the principle of caution be applied as regards exposure to magnetic fields;
- all unnecessary exposure be avoided;
- new places of work be designed and equipped in such a way that the exposure to magnetic fields is minimized;
- the manufacturers of electrical equipment aim to minimize the magnetic fields;
- manufacturers give details of the levels of magnetic fields in connection with the sale of such equipment;
- no employee be exposed to an average exposure exceeding 2 milligauss per working day;
- temporary high exposures be minimized as far as possible;
- the employer map out the existing levels of magnetic fields and, when necessary, draw up plans of action in accordance with the internal control regulations;
- the staff in question be informed and trained;

- practical measures to reduce exposure be taken without further delay, such as indication of areas with high exposure, reduction of magnetic fields, transfer of work sites, and changed work organization;
- the distribution of responsibilities between the different authorities be clarified;
- Swedish laws and ordinances reflect a viewpoint corresponding to the principle of caution;
- stray currents be eliminated by the introduction of five-wire systems (three hot phases, one neutral, and one grounding conductor);
- the National Board of Occupational Safety and Health, pending the draft for a hygienic limit, issue a regulation in accordance with the views already mentioned;
- and that the research on electric and magnetic fields be continued and intensified.

American unions have tried to get their government to assess workers' exposures and set occupational standards for magnetic fields, but have met with little success. In the late 1970s, some important occupational research, centered mostly around VDT exposures, was being conducted by the government, but the Reagan administration stopped it. Some interest is again being shown, however.

Many European countries have regulations for electric fields, but only a few have set standards for magnetic fields. Often what is allowed is ridiculously high and provides no protection at all. Germany and Poland allow 5 gauss for power-utility workers; Russia allows 7.5 gauss for one-hour exposures and 1.8 gauss for an eight-hour day. England has set limits of 1.7 gauss for utility workers. Consider these levels in light of Mantanowski's work, which found telephone workers to have seven times the cancer risk of other professions at exposures of only 4.3 milligauss—a thousand times lower. And consider that electrical workers, after being exposed to high magnetic fields all day, return home to the average background exposures, too.

Radio-Frequency/Microwave Professions

Far more bioeffects research exists for the 50- to 60-hertz frequencies than for the radio-frequency and microwave (RF/MW) bands, but what does exist is troubling. Many would say that the absence of studies is not an accident, that important research has been systematically ignored, blocked, discredited, or otherwise undermined by those who stand to lose the most—the U.S. military and various segments of industry. Many also now think that these frequencies, which are ubiquitous today, will prove to be as detrimental as the power-line frequencies, if not more so. That is because the human anatomy absorbs radiation most efficiently in the FM radio band, although the other RF/MW bands are harmful as well.

The RF/MW professions include broadcast and radio engineers and employees at radio/TV transmission facilities; those who work in multimedia environments; ham-radio operators; civilian and military air traffic controllers; civilian and military radar technicians and law enforcement officers who use radar guns; those who work for paging facilities; and operators of RF heaters and sealers (shrink-wrappers).

Segments of the medical profession also increasingly use RF/MW technologies, including diathermy therapists and MRI technicians. Airline personnel, especially pilots and flight attendants, have high RF/MW exposures, partly from the avionics equipment on planes but also from the convergence of the many frequencies being transmitted from the ground into the higher altitudes. Just flying in a plane will subject a person to higher exposures, especially higher cosmic rays—something frequent-flier passengers should be aware of.

The standard most widely used for RF/MW exposures is the ANSI/IEEE C95.1 standard, which has been adopted in its various revisions by the FCC, OSHA, and some states. But, as we have said, the standard is based on thermal-effects-only and is considered inadequate for nonthermal bioeffects. (See

Chapter 2 for the history of ANSI/IEEE.) We are effectively allowing increasing amounts of RF/MW-generating technologies to proliferate, with no adequate protection for either the military or the civilian population. Those who have been watching the controversy for decades feel that the ANSI/IEEE standard may soon crumble, largely because of ANSI's adherence to thermal effects as being the only possible bioeffect. The standard has fallen into such disrepute that the EPA and the National Institute for Occupational Safety and Health (NIOSH) have asked the FCC not to adopt its most recent version. Some air force bases have actually been setting their own standards, as have several states (Massachusetts is one) and municipalities (Portland, Oregon, and Seattle, Washington).

Here are some of the studies relating to the RF/MW frequencies:

- McLaughlin reported in 1953 internal bleeding, leukemia, cataracts, headaches, brain tumors, heart conditions, and jaundice in radar workers.
- Johns Hopkins School of Medicine researchers found in 1964 correlations in the incidence of Down's syndrome in children whose fathers worked near radar sites.
- Peacock in 1971 found a surge in birth defects (clubfoot especially) in the children of radar-exposed Army helicopter pilots.
- Stodolnik-Baranska reported in 1973 that human lymphocytes exposed to pulsed microwaves showed several abnormalities that were directly related to the length of exposure.
- Friedman found in 1981 that radar technicians were twelve times more likely to get polycythemia (a rare blood disease characterized by too many red blood cells) than the rest of the population.
- Milham found in 1985 that ham-radio operators had an increased leukemia mortality and a greatly increased incidence of other cancers, including Hodgkin's disease and lymphomas.

- Szmigielski reported in 1985 a significant increase in all cancers in Polish career military personnel who were exposed to high RF/MW. Findings included a three times higher likelihood of developing cancer than among nonexposed servicemen, with increases in tumors of the blood-forming organs and lymphatic tissue at nearly seven times that of non-exposed personnel. Increases in melanomas were also seen by Szmigielski, Vagero, and Tornqvist. Strong correlations in length of exposure and cancer formation were noted. Szmigielski in previous work had found that chronic microwave exposures increased the incidence of high blood pressure, headaches, memory loss, and brain damage. He is also known for his studies of microwaves and the immune system, which found a biphasic immune-system reaction to RF/MW exposures—the whole immune system was stimulated initially, followed by the gradual, transient suppression of all immunity after increasing periods of exposure. (These are some of the symptoms of chronic fatigue syndrome and other immune system disorders.)
- Swedish researchers found in 1985 changes in cerebrospinal fluid in radar workers.
- Henderson and Anderson reported in 1986 significantly elevated cancer rates in men and women living near broadcast towers in Honolulu, as well as increases in childhood leukemia.
- Lester and Moore reported in 1984 that neighborhoods exposed to radar waves from two airports in Wichita, Kansas, had a higher cancer incidence than nonexposed neighborhoods. The cancer incidence was highest in areas exposed to radar from both airports. The authors also studied areas near air force bases with radar equipment and found a significantly higher cancer incidence and mortality in such locations, too. (Although these last two studies are of residential exposures, an occupational extrapolation can be assumed.)
- Cleary, Li-Ming, and Merchant in 1990 found that RF/MW

radiation stimulated human brain-tumor cells at low levels but suppressed it at higher levels. The effects continued for up to five days after an exposure of only two hours.

- Goldoni found in 1990 that air traffic control workers exposed to radar had a higher incidence of posterior cataracts, as well as changes in blood chemistry and brain activity.

- Garaj-Vrhovac, Horvat, and Koren reported in 1991 that workers exposed to radar radiation showed particular kinds of chromosomal abnormalities. Additional studies of cell cultures of both human lymphocytes and hamster cells showed the same chromosomal aberrations, as well as the formation of micronuclei (chromosome fragments that indicate genetic damage).

- Kues and Monahan reported in 1991 that laboratory experiments with monkeys and clinical studies of workers accidentally exposed to weak microwave radiation showed damage to the eye's cone photoreceptors and rods. (Cone damage can result in a loss of color discrimination and decreased visual acuity; rod damage can affect night vision.) In other work, the researchers found that low-level pulsed radiation could damage the endothelial layer of the cornea and cause leakage in the eye's blood-aqueous barrier. The researchers have also found that glaucoma medications can make the eyes much more sensitive to microwaves at levels far below what have been considered safe.

- Tynes reported in 1994 that women working as radio and telegraph operators for more than nine years (mostly aboard ships) were 80 percent more likely to develop breast cancer than nonexposed women. The risk increased the longer the woman worked; after twenty years, the risk rose to 2.2 times the expected rate. (This was the first study to link RF/MW exposures with breast cancer, although five studies have done so with power-line frequencies.)

In addition, hundreds of animal studies of the RF/MW frequencies, some dating back to the 1920s, have important implications for human exposure. Here are only a handful:

- Prausnitz and Susskind reported in 1962 that a third of test mice developed leukemia when exposed to RF frequencies like those from air force radar transmitters.
- Frey reported in 1975 changes in the blood-brain barrier in rats exposed to pulsed microwaves. In earlier work, Frey also found that microwaves could speed up, slow down, or stop altogether the heart rhythm of frogs when the pulse was synchronized with the heartbeat.
- Army scientists reported in 1977 a leakage in the blood-brain barrier (which controls toxins' admittance into the brain) from pulsed microwave frequencies in test animals.
- Yao reported in 1978 chromosome alterations in the corneal cells of Chinese hamsters exposed to microwaves.
- Takashima reported in 1979 that rabbits exposed to RF radiation showed changes in brain waves.
- Chang reported in 1980 that mouse leukemia cells exposed to pulsed and static microwaves showed an inhibited synthesis of DNA.
- Guy reported in 1985 that several generations of rats exposed to pulsed microwaves in ranges that simulated the levels allowed by current standards for humans had increases in adrenal medulla tumors, malignant endocrine and ectocrine tumors, and increases in carcinomas and sarcomas.
- Balcer-Kubczek and Harrison reported in 1989 that microwave exposures of mouse embryos both initiated tumors and interfered with immune-system competency.
- Salford and Persson reported in 1992 that increases in the permeability of the blood-brain barrier of rats could be effected at extremely low levels of microwave radiation of around 915 megahertz (around the frequencies used by cellular phones and cellular-phone towers).
- Lai and Singh reported in 1994 that a single two-hour

exposure to 2.45 gigahertz radiation could increase single-strand breaks in the DNA of brain cells in rats.
• Sarkar reported in 1994 that the DNA of brain and testicular tissue in mice showed rearrangement following microwave exposure at 2.45 gigahertz at intensity levels currently considered safe.

Some studies revealed a higher cancer incidence in those exposed to several frequency ranges. Pearce found an excess leukemia risk in EMF occupations, with the highest incidence among radio and TV repair workers. Thomas found in 1987 that workers with both RF and ELF exposures had double the risk of developing brain tumors than the general population. And several studies have shown a greatly increased tumor incidence when chemical and EMF exposures were combined—verifying the work done on EMFs as co-teratogens.

In addition, cancer clusters (of Hodgkin's and of breast and cervical cancers) and abnormally high rates of ALS (a degenerative nervous-system disorder) have been found in Florida neighborhoods located near FAA and air force radar sites. The same kind of radar unit is suspected in a brain-tumor cluster found among U.S. Census Bureau employees in Suitland, Maryland. (A 1986 study also found an increased incidence of ALS among those who worked in electrical professions.) Other cancer clusters have been found near communications towers in Honolulu and Lualualei, Hawaii, in McFarland and San Francisco, California, in Portland, Oregon, and in Thurso, Scotland (an increase in childhood leukemia).

A Note about Radar

Not all radar is alike. The radar in the towers used for air traffic control is not the same as that used by the military to track satellites or communicate with submarines, or the same as Doppler radar, which tracks weather patterns. Although all these forms are within the microwave frequencies, they use more or less power, depending on their purpose and the distance they need to cover, and so the exposures from them will vary. Some are also modulated at different frequencies, meaning that two frequency waves are used together, and more bioeffects have been observed with modulation.

New radar systems and technologies are coming on-line every day around the globe, and they are being met with increasing opposition. Phased-array radar, which is considerably stronger and has sharply higher exposures over the older scanning systems, is meeting with opposition in England. NEXRAD systems for the National Oceanic and Atmospheric Administration, 173 of which are planned worldwide, with approximately 10 already on-line, are also meeting with opposition.

Hand-held radar guns, like those used by traffic-control officers, and security systems, like those around shops and malls, are within the microwave range, but they operate at far less power than any of the systems just mentioned. Nevertheless, problems have come up with radar guns, especially among police officers who cradled them in their laps while the guns were operating. There are upward of twenty lawsuits in several states concerning adverse health effects on law-enforcement officers who used radar guns. The suits involve skin cancers (melanoma, basal and squamous-cell carcinomas, and liposarcomas), lymphomas (non-Hodgkin's, multiple myelomas), leukemia, and thyroid and testicular cancers.

The police radar guns operate with a continuous wave signal at either 10 gigahertz (called the X-band) or 24 gigahertz (called the K-band). They are both high enough to directly resonate some biological molecules, meaning that they may have a direct effect on the body even though they operate at very low power. The X-band models have an average power density around their antennas of about 3.36 mW/cm^2 for fixed mounted models and 2.66

mW/cm² for hand-held models; K-band models average lower, around 0.93 mW/cm² for fixed mounted systems and around 0.69 mW/cm² for hand-held devices.

While these levels are below the ANSI/IEEE guidelines for thermal effects, they are nevertheless a million times higher than some field strengths that have been shown to alter cell growth with direct exposure. German studies within this frequency range using yeast cells (a common cell type often used in biological and biomedical research on cell growth and genetic mechanisms) found that cell growth was turned on or off in a very specific fashion, as if the cells were tuned to weak microwave radiation in a frequency-selective way. Subsequent research indicated that lower power actually increased the tuning effect. In other words, there may be no safe level of exposure for some microwave frequencies.

The practice of law-enforcement officers of resting an operating radar gun on the lap is of particular concern, since a majority of the cancers reported have been in the testicles or have been lymphomas originating in the groin. Most of the energy penetrates the body and becomes absorbed in tissues, and this is generally thought to be dissipated (attenuated) fairly quickly. Areas of the body with fat or muscle layers may provide more protection. But around the groin, the lymph glands and the testicles are extremely close to the surface and can indeed be directly affected. The state of California has, in fact, found that police officers have a 2.68 increased risk of death from lymphatic cancer—double that from other cancers.

In addition, police officers have RF exposures from their two-way radios (which also may be carried on the body with the antenna close to the head). And motorcycle officers often have antennas mounted behind them, which effectively places their bodies in the middle of an RF field. Squad cars with antennas mounted on the back have the same effect—to say nothing of the possibility that exhaust fumes and EMFs are acting as coteratogens.

If you are in law enforcement and use a radar gun, mount it away from your body, preferably outside the vehicle but not right next to your head. (The FDA recommends at least six inches from any part of the body.) Do not allow the antenna to point at you when the gun is in operation. Some police departments are giving officers a choice of whether to use the guns or not, and the new laser models becoming available may be better for the officer (but not for the oncoming driver being tracked).

Other RF/MW Professions

Several other occupations make significant use of EMF/RF/MW technologies for various periods of time throughout the day. Photographers and graphic artists leaning over lightboards can bring their heads close to 60-hertz magnetic fields averaging around 80 milligauss all day long. Emergency medical service (EMS) technicians and any others who use two-way radios regularly will have high intermittent RF exposures. There have been reports of menstrual-cycle irregularities in female EMS workers after their older radios, which operated at 400 megahertz, were replaced by newer models operating at 800 megahertz in the microwave/cellular-phone range. (Two-way radios use more power than cellular phones.)

But by far the greatest concern is for those who work in heat-sealing operations. Everything we purchase today that is sealed in plastic—which is just about everything—or laminated has been sealed by a worker using an RF heating device. Plastic molders and sealers, of whom there are many thousands—usually in low-paid positions with few medical benefits, the vast majority of them women of childbearing age—can be exposed to RF levels far above even the old ANSI/IEEE standard, sometimes over 260,000 microwatts, according to one NIOSH survey. No health studies have been done of this occupational group, but one might expect that miscarriages and birth defects would be elevated, as well as glandular, blood, and central-nervous-system cancers.

It would cost very little to make most RF-heating machines exposure-free. It could be done much like sealing the leaks around a microwave oven.

Ham and CB Radios

Most operators of ham radios truly love their hobby, and those who rely on CB radios would not give them up for anything. But these are RF-transmitting devices, and operators should be careful where they position the antennas.

Never have the transmitting antenna near you. Sometimes an operator sets up a high-intensity antenna in the same room, or an amateur radio operator will clip an antenna right to the shoulder. Some mobile operators mount antennas on their cars right next to their heads. Distance is very important. It is best to get an appropriate meter and measure to find out where the fields are.

CB radios should not be placed at genital level in the vehicle. The glove compartment is better, unless you have frequent passengers.

A few lawsuits for "electronic trespass" are being filed against ham-radio operators. Although mounting an antenna outside is certainly better for you, it might not be better for your neighbors.

Medical Professions

One of the great ironies about EMFs is that mainstream medicine's abandonment of energy modalities in the health paradigm is coming back to haunt the profession in several ways. One is through electromagnetic interference between the machines themselves, which can cause a life-threatening situation if the equipment fails. The other is through direct health effects on medical personnel. (If all that equipment is interfering with other machines, imagine what it is doing to the humans in its path.)

High-tech medicine today is truly a high-risk profession for EMFs. Surgeons and operating-room technicians spend hours surrounded by a variety of equipment that generates many frequencies at different strengths, as does the equipment in neonatal units. Diagnostic imaging, once strictly confined to X-rays (which can be shielded by lead aprons), has been joined by ultrasound (in the RF range) and magnetic resonance imaging, which uses powerful DC magnetic fields and RF frequencies.

Medical professionals are beginning to show adverse reactions. Operating-room personnel are becoming sensitized to certain machines. Some have had to switch to other jobs within the medical field or to leave the profession altogether.

One study by Drs. Josephine Evans and David Savitz, published in the *Journal of Occupational Medicine*, found a threefold increase in spontaneous abortions in MRI operators, although infertility and the incidence of low birth weight were not increased. Also, in a Johns Hopkins School of Hygiene and Public Health study, physical therapists who used microwave diathermy equipment on patients during or shortly before their first trimester of pregnancy had a 28 percent higher risk of miscarriage, although those who used shortwave diathermy units did not. The miscarriage risk doubled as the number of sessions increased. (It is thought that MW radiation is better absorbed by watery tissue like amniotic fluid than shortwave radiation is.)

Other studies, however, have linked shortwave diathermy with reproductive problems. Danish researchers, led by Dr. Anders Larsen, reported that female shortwave-diathermy therapists gave birth to fewer boys, who also tended to have low birth weights. And Swedish researchers, led by Dr. Bengt Kallen, found an increase in stillbirths and birth defects among such therapists. Also, Dr. Stanford Hamburger at the FDA reported a significant link between male shortwave-diathermy therapists and heart disease.

Leakage from diathermy units can be substantial and can vary among manufacturers. Measure with the appropriate equipment and contact the manufacturer if you find readings over 1 mW/cm^2.

Electromagnetic Interference

Electromagnetic interference (EMI) is a growing problem, especially for the airlines, which have had to ban the use of personal radio transmitters on most flights, and for the medical professions, which are only beginning to discover what EMI can do.

There are over a hundred reports, as of this writing, of episodes of hospital equipment interfering with other equipment— infusion pumps affecting patient monitors, muscle stimulators causing therapeutic tables to move without warning, and disruptions in the operation of fetal heart monitors, incubators, dialysis machines, and ventilators caused by cellular phones or two-way radios. Several deaths have been attributed to EMI.

Apnea monitors, anesthetic gas monitors used during surgery, and electric wheelchairs are of special concern. Some electric wheelchairs have actually been driven off curbs or piers by signals from police or fire equipment, harbor control boats, or CB or ham radios. The FDA's Center for Devices and Radiological Health has issued recommendations for electric wheelchair manufacturers to achieve a certain protective level against EMI, to state what the level is on new chairs, and to include a warning label about EMI on all new chairs. For the wheelchair user, the FDA recommends turning off the wheelchair before using a cellular phone and being alert to nearby transmitters.

Swiss and Italian studies have found that cardiac pacemakers can be disrupted by EMI from digital cellular phones (the newer models just coming on the market, not the older analog models). Pacemakers were both speeded up and slowed down when some phones were brought close to the patient's chest. Apparently the pacemaker misidentifies some of the digital signals as the heart-beat, especially an 11-hertz pulsed signal.

Hearing aids are also malfunctioning because of EMI. Cellular phones operating on the European standard can produce 200-hertz tones in hearing aids from a distance of 30 meters; at close range, the sound in the wearer's ear may be as loud as 130 decibels. American phone models can also interfere with hearing aids. And some elderly patients have been misdiagnosed with auditory hallucinations when, in fact, their hearing aids have corroded in a way that allows them to pick up various broadcast signals.

As the ambient electromagnetic background intensifies from the increasing number of wireless devices, the situation can only be exacerbated. One FDA scientist recommends that hospitals ban all cellular phones and two-way radios in their vicinity, as well as rooftop commercial transmitters of any sort. (Hospitals often lease roof space to telecommunications companies.) There is also speculation that many of the recent unexplained airplane crashes will eventually be attributed to EMI.

It is still left up to individual manufacturers of various medical devices to figure out how to reduce emissions, but not many of them know enough about the problem. Shielding is possible for most equipment. Concern over the functioning of the equipment, rather than bioeffects, may be the reason for improvement, but humans will gain in the process, too.

Chapter 14

POWER LINES

THERE HAS BEEN tremendous public concern about high-tension power lines in the last several years. Property owners in the vicinity of high-tension lines have worried about their health and that of their children, as well as watching the value of their homes plummet. Numerous citizens' groups have formed to block new power-line corridors and substations, and scores of lawsuits have been filed claiming adverse health affects or seeking monetary compensation for property devaluation.

The utility companies, once contemptuous of bioeffects research, have come around to a "maybe there's something to it, but we need more data" approach (although their legal stance toward consumers is still "prove it"). Many consumer advocates see this as a new variation on an old obstructionist theme; others see in it an important if subtle shift and a sign of progress.

Consumers must understand that literally billions of dollars may be at stake for the utilities. They are businesses, which must answer to their stockholders—although in a more limited way than other businesses, since most utility price structures (and therefore stock dividends) are government regulated. It is equally important to realize that the historical mind-set of the utilities has been that of the "good guy." They have always had an acute sense of their public responsibility to keep the power on—to prevent life as we know it from coming to a screeching halt in a power failure, which can be life-threatening for hospitals, for the elderly or the homebound infirm, and for those living in extremely cold climates. So imagine the consternation of the industry to suddenly find itself cast as public enemy number one—just shy of being baby killers, in the minds of some.

This is not to say that the utilities have been shining examples of concern for consumers when it comes to the bioeffects of electromagnetic fields, nor to deny that they have dragged their collective feet in making changes. It is only to lend perspective to the present situation, to diffuse the polarity between an emerging scientific understanding and the monumental changes that may be required of the utility companies (including the behavior of some executives, who have been not only defensive toward inquiring consumers but self-congratulatory when they win legal judgments).

The question is, is there a middle ground between consumers entitled to safety and a utilities industry that has brought us power for nearly a century and would like to continue doing so? Or is our only alternative to turn back the clock to a preindustrial world? There is a middle ground, and that's what this chapter is partly about.

How the Electric-Utility System Works

Power reaches our homes and workplaces through what is called the transmission and distribution system. The transmission part of the network includes the huge generating plants that gather power through hydroelectric processes, such as at Niagara Falls, through the burning of fossil fuels like oil or coal, or through nuclear fission at nuclear-power plants. The voltage from these plants is immediately increased in step-up transformers and is moved along high-voltage transmission lines strung on tall, specially designed towers. When it reaches the community where it is needed, it is decreased in step-down transformers in substations. It can then be used in local households and businesses through the distribution end of the network.

The power generators, high-voltage transformers, substations, and transmission lines of the transmission network deal with the high-voltage side. The distribution side is made up of substations and the poles and lines that run through our neighborhoods. The distribution lines generally use lower current than the large transmission lines, although there are exceptions.

The distribution network is broken down into primary lines and secondary lines. Primaries use intermediate voltages to serve a whole region. The primary lines are mounted at the tops of poles, are thicker, and are connected by insulators. They carry current (4 to 69 kilovolts) to smaller step-down transformers, which may be pole mounted, encased in small metal structures at ground level, or sometimes buried. Step-down transformers reduce the voltage of the primary lines so that it will be usable in domestic wiring; otherwise there would be danger of shock, as well as insulation problems. Secondary lines and domestic current operate between 115 and 240 volts. Secondary lines are located slightly lower on the poles, and it is these that bring the current into your home at the service drop. Beyond the service drop, the customer end of the network takes over. The illustration "Electrical Transmission and Distribution System" depicts the entire network.

Transmission lines and many distribution lines use three "hot" wires, with voltages that oscillate at 60 hertz although in different phases. (Rephasing can reduce magnetic fields and will be discussed later.)

While consumers have focused their greatest ire on the giant towers that crisscross the nation, many do not realize that the local distribution wires in front of their homes can—at peak times of day—carry just as much current as the huge transmission lines. That is one reason for taking measurements in the rooms nearest to the street as well as at the service drop

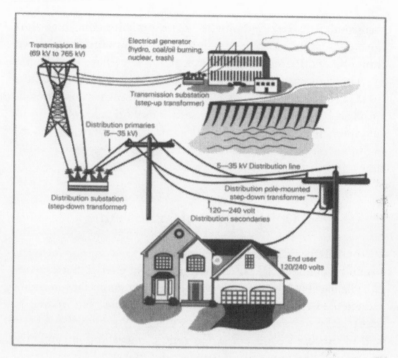

Electrical Transmission and Distribution System

(where the wires connect to your building) at different times of the day.

In addition, substations are often located very close to residences. Sometimes they are difficult to recognize, if they are installed in large, rather plain-looking buildings. Outdoors, they are readily identified by their characteristic rows of coils, wires, and insulators. Nothing else looks quite like them.

Cancer clusters have been reported around some substations that are especially close to residences. Electromagnetic fields are often quite strong near substations because of the high-current lines coming down into them, bringing the lines that much nearer the ground. Also, some homes have been identified as high current, depending on their proximity to pole-mounted step-down transformers. Increases in a number of cancers have been found in residences near high-tension lines and in high-current homes.

There are currently 350,000 miles of transmission lines in the United States, and approximately 2 million miles of distribution lines. It is a massive network. Virtually everyone is exposed to 60-hertz fields. There is no such thing as a nonexposed population—it is just a question of degree.

A Quick Review of Some Terms

At this point, a quick review of some scientific terms is needed. Voltage is the potential of electrical energy in a line (recall the garden-hose analogy), while current is the movement of electrical charge through the line (when the hose is turned on). Electrical potential is measured in volts (V) or kilovolts (kV); 1 kilovolt equals 1,000 volts. Current is measured in amperes (amps) and the power used is measured in watts (W). Watts are calculated by multiplying the current by the voltage.

In America, electricity operates at 60 hertz (Hz) of alternating current (AC), which is in the extremely low frequency (ELF) range. (A hertz is the measurement of the rate of fluctuation of the current.) Alternating current changes direction, unlike direct current (DC), which flows in a single direction. (Batteries are a common example of DC power.) Sixty hertz of alternating current reverses itself 120 times a second, making 60 complete cycles per second.

Electric fields and magnetic fields exist in space. Electric fields are present any time there is voltage. They are measured in volts per meter (V/m) or, when in large units such as those around high-tension lines, kilovolts per meter (kV/m). Magnetic fields are created around (perpendicular to) electric fields whenever current is flowing. Magnetic fields are measured in gauss (G) or tesla (T); 1 tesla equals 10,000 gauss. Because these measurements can be large, smaller units are usually used, like milligauss (mG) and microtesla (μT); 1,000 milligauss equals 1 gauss, and 1 microtesla equals 10 milligauss.

Electricity moves along pathways, or conductors, that allow electrical charges to flow through them. Conductors include certain kinds of metal wires, the human body, and some substances, such as water. Some conductors are more efficient than others. Nonconductors (such as plastic) and insulators do not allow electricity to flow through. An electrical circuit is a loop controlled by a switch that can interrupt the flow of electricity. A lamp, for example, turns off when the switch introduces a space between two contacts, which then will prevent the current from flowing freely through the pathway.

Everything in the 60-hertz system (50 hertz in Europe) is contained in a circuit—from the reading light you use to the entire transmission and distribution network that brings the capability into your home.

Laypeople often confuse the energy created around high-tension lines with the radio-frequency radiation near broadcast towers, but the two behave quite differently. Radio-frequency radiation "radiates" away from its source; the energy around high-tension lines "decays" rapidly with distance. (There are other significant differences between them as well. For more information, see Chapter 3.)

High-Tension Lines

Many sizes and shapes of towers and poles are used to support high-tension lines, based on what voltage is driving the current over how long a distance. Transmission-line towers are an indication of the voltage being carried—the higher the voltage, the taller the tower, and the wider apart the conductors (wires) must be spaced. Many consumers find them frightening in their

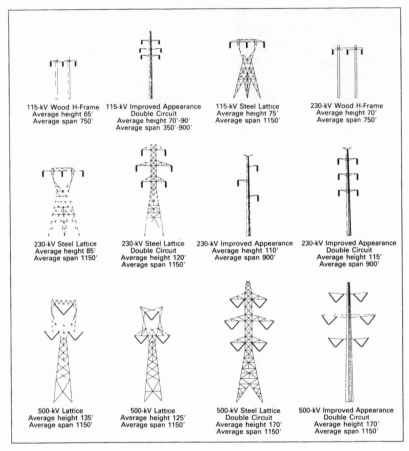

Some Common Tower Styles

USED WITH PERMISSION FROM THE BONNEVILLE POWER ADMINISTRATION.

immensity, but some writers have discovered a kind of grotesque beauty in their industrial symmetry. (The illustration shows some common types of towers.)

Most transmission lines in the United States are alternating current and use three electrical phases that function through arrays, or bundles, of wires. Each phase carries power that peaks at a slightly different time. As the voltage on one wire is peaking, the voltage on one of the others is one-third of a cycle ahead and the voltage on the other is one-third of a cycle behind. (This is generally referred to as the three phases of the power network.)

Fields that are exactly in phase with each other—meaning they rise and fall together—can create stronger EMFs. But phases that are timed to be exactly opposite each other, meaning that one is rising while the other is falling, will result in a much weaker field, often canceling out the field altogether. (Phasing alterations as a utility mitigation technique will be discussed later in this chapter.)

Transmission lines and many distribution lines use this three-phase power. In addition, many commercial and industrial facilities require three-phase power to run machines that draw a lot of current. But usually the 120/240 volts of domestic power are supplied by a single phase from the primary line, which produces two "hot" lines and a neutral one for the home. (The voltage between the hot lines is 240 volts, and between either hot line and grounded neutral is 120 volts.) Utility companies try to service equal numbers of homes with the same phase in order to balance the load.

High-tension lines carry different amounts of voltage and current. This can vary from 69 kilovolts all the way up to 750 kilovolts. (There are also some experimental 1,200-kilovolt lines.) Usually the higher the voltage, the taller the tower must be. In addition, there are several different ways of hanging the cable. Different cabling configurations will produce significantly higher or lower magnetic fields.

In addition to the AC transmission lines, there are about five overhead DC lines in western America and approximately ten worldwide. DC lines create nonfluctuating magnetic fields, but they are practical only for very long runs, because it is expensive to tap off of them along the way since conversion equipment between AC/DC systems is costly.

Power is moved on transmission lines at a very high voltage in order to reduce energy losses along the way (the higher the voltage, the lower the losses). Energy loss can be as high as 40 percent, according to some estimates for AC lines, because of the constant reversal of energy. DC lines have far less loss because the current flows in only one direction. DC lines are also cheaper to construct because they require only two lines rather than three, smaller towers, fewer insulators, and smaller rights-of-way.

The utility companies continue to experiment with new ways of hybridizing DC with the AC systems already in place. On the face of it, it seems like a good idea. Caution is in order, however, because there is some indication that AC magnetic fields in combination with certain DC fields may be causing adverse bioeffects.

The Electromagnetic Environment Around High-Tension Lines

Anyone who lives near a high-tension corridor knows that the lines can do strange things. They snap, crackle, pop, hiss, and hum; they sometimes create arcs of visible electric charges; and fluorescent light tubes will glow underneath them without being plugged into a power supply other than the human hand. (Fluorescent tubes will do the same around any strong electric field, including operating engines and CB radio antennas.) In addition, the air around high-tension lines can be electrically charged in a way that will, among other things, make the hair on your arms and head stand up.

There are also reports of farmers being zapped on metal roofs near high-tension lines, and of a phenomenon called stray voltage—when electrical current flows through the earth into an area where it doesn't belong and creates constant small electrical shocks in livestock, such as a milking cow standing on a metal surface. In such cases, high rates of miscarriage in cows and birth defects in their offspring, as well as increased morbidity and mortality, have been observed.

The electromagnetic environment around high-tension lines can sometimes seem like living under a thunderstorm that produces neither rain nor thunder and is in full sunlight. It is truly an altered, artificial environment. These are some of the physical properties of such an electrically charged atmosphere:

- Corona is an effect that occurs on insulators, conductors (wires or lines), and hardware in areas where there is a high electric-field strength, and it involves the partial breakdown (or ionization) of air particles. Corona can result in TV and radio interference, audible noise, visible light, and the production of small amounts of oxidants like ozone and nitrous oxides. Corona discharge increases in higher altitudes because the strength of the air particles is less than at sea level. Bad weather (rain, fog, snow, and condensation) all increase corona.

  The audible noise associated with corona is usually caused by water droplets on the wire's surface. Noise near some AC lines can produce a 120-hertz hum, and near 500-kilovolt transmission lines the levels can be as high as 60 decibels. (Vacuum cleaners average around 70 decibels, garbage disposals around 80.) The actual nuisance factor depends on other background sounds. Corona noise will seem higher in rural areas than in louder urban settings. Utility companies have reduced corona and the accompanying noise through more efficient designs, but it is not possible to escape it completely. Anyone who has stood near a high-tension line or has had radio interference while driving under one has experienced corona effects.

- Air ions are air particles that often form near high-tension lines. They are naturally occurring molecules with either extra electrons (negative ions) or missing electrons (positive ions). They are formed through the collision of particles stirred up by storms, sunlight, a waterfall, radioactive materials, wind, cosmic rays, and electrical corona.

Air ions occur in extremely small concentrations and differ in rural and urban environments. Near AC transmission lines, positive and negative ions constantly form but rarely extend much beyond the right-of-way because the positive and negative ions attract each other. The situation is more complex around some of the high-transmission DC lines in the west. These DC lines are in a constant state of corona, and ions can be carried by the wind far beyond rights-of-way. The illustration shows the complex ion environment around a DC tower.

Some have claimed that negative ions have a therapeutic effect on people who are sensitive to them. Negative-ion generators have been marketed commercially for some medical problems, like burns, for the relief of allergies, and to create

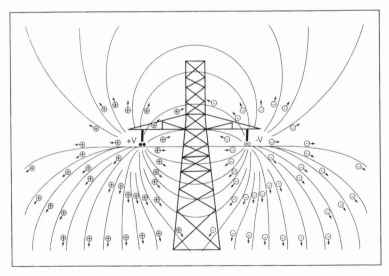

Complex Air-Ion Environment Around a High-Tension DC Tower
ADAPTED FROM THE BONNEVILLE POWER ADMINISTRATION'S
ELECTRICAL AND BIOLOGICAL EFFECTS OF TRANSMISSION LINES, A REVIEW.

a healthier home or office environment. The so-called sick-building syndrome is said to be partially caused by an absence of negative ions. But many people report discomfort in the presence of negative-ion generators; especially prevalent are respiratory and neurological complaints.

A body of contradictory studies exists on the effects of air ions. One hypothesis (by Krueger and Sigel) holds that when positive ions are inhaled, they cause the release of seratonin in the brain. Negative ions, on the other hand, speed up the removal of seratonin. Another group of researchers (Kavet and colleagues) have proposed that such interactions are mediated through specific neuroendocrine cells found in respiratory tissue. There are over thirty-three air-ion studies, reporting on a variety of behavioral and medical responses in test animals and humans. But no can can say with any certainty whether this one aspect of the ambient environment around high-tension lines causes ill effects in humans. Reported effects have been mild and apparently transient, similar to the reactions of people who are sensitive to weather changes when air-ion alterations also take place. Nevertheless, more studies are called for; changes in neurohormone levels are nothing to dismiss.

How Close Is Too Close?

Many people now think that the normal distance between power lines and populated areas is not great enough. Several municipalities and states have moved to increase the distance from rights-of-way, especially when schools or other structures are near a new line under construction. Obviously it is easier to do this before the line goes in than afterward.

Established power lines are more of a problem. Many older lines were built when areas were rural, but now development has gradually encroached on them. Also, distances were originally set mainly to protect against shock from electric fields; magnetic fields were not considered as a human or environmental risk until recently.

Several knowledgeable scientists have said publicly that they would not live near high-tension lines. But if you do live near one, it's important to know that the risks, while documented, are relatively small in comparison to other risks completely within your control—such as smoking. (For children or a woman planning to become pregnant, the risk becomes a little greater.)

Most homes that have high ambient magnetic-field readings (much above 3 milligauss) either have wiring problems or are close to power lines. A utility company, upon request, will take measurements throughout such a home, and magnetic-field measurements are becoming standard information in real-estate transactions in many areas of the country, not unlike radon and water-purity tests.

For aesthetic reasons, real estate has always been less expensive near high-tension lines, and this inexpensive land has too often tempted municipalities to erect schools near them. Inexpensive land has also attracted real-estate speculators, who have developed whole neighborhoods as close to the edge of rights-of-way as was legally allowed, in some cases less than fifty feet. Most of this happened over the last ten to twenty years, before the public became better educated and some municipalities started requiring much larger setbacks.

But what to do about homes, schools, and businesses already located near power lines? That is a difficult question for the public and the utilities alike. Prudent avoidance in such cases is largely up to the utility company and involves buying larger tracts of land and burying, rephasing, or reconfiguring the lines. There is little an individual homeowner can do except move out.

But first make sure you know what you are dealing with. If you do not trust the utility company to take accurate measurements, hire an independent consulting company or purchase your own gaussmeter. Move your children's bedrooms as far away as possible from the lines or from transformers. Take readings around the head of each bed in particular, and be sure to measure any room in which a lot of time is spent. Schools can sometimes be rearranged so that classrooms are as far away from the lines as possible; with former classrooms becoming storage space. Increasingly, parents are refusing to send their children to—and teachers are refusing to teach in—schools located near high-tension lines.

Utilities tend to think this is mass hysteria generated by the media, but only you can decide what level of risk you are willing to take. Many people take risks with themselves that they will not take for their children. Unfortunately those who try to sell homes near high-tension lines often find they cannot attract buyers, even at substantial reductions. There have been lawsuits against utilities for "illegal takings" with varying results. The legal data base on the subject is growing, however. If you want to follow this route, try to find an attorney who has already handled EMF suits. An attorney who simply specializes in real-estate law would have to spend too much time reading up on a vast and controversial subject—and utilities lawyers already have a tremendous amount of savvy in defending such cases.

Magnetic Fields Near High-Tension Rights-of-Way

In the following table are some typical readings near different types of power lines. The actual field strengths and the width of rights-of-way vary, depending on the line's design and the voltage being carried, as well as the amperage, which changes according to demand. These are just averages of the fields, measured at various distances from the source.

In addition, there are many lines functioning at 345 kilovolts with rights-of-way of about 140 feet. The magnetic field at the edge of these rights-of-way averages about 100 milligauss at 50 feet.

Line Voltage	Typical Right-of-Way Widths	Electric Field (kV/m)				Peak Magnetic Field (mG)			
		50–65'	100'	200'	300'	50–65'	100'	200'	300'
115 kV	50'	0.5	0.07	0.01	0.003	10	2	0.6	0.3
230 kV	90–125'	1.5	0.3	0.05	0.01	30	10	2	1
500 kV	135'	3.0	1.0	0.3	0.1	50	25	7	3

Figures are from the Bonneville Power Administration Review, 1989.

Some question these measurements, saying they are much too low, especially where existing lines have been upgraded to carry more current. One transmission-line coordinator for the state of Florida found 6-milligauss magnetic fields as far as 2,000 feet from a 500-kilovolt line.

Many people hike, ride motorcycles or dirt bikes, or cross-country ski under high-tension lines because of the long, clear, flat stretches. Some municipalities have located school playgrounds and athletic fields within rights-of-way. The magnetic fields directly under any power line, especially midway between two towers, where the lines dip the lowest, will be high, and such activities are not recommended, even for short periods of time. Magnetic fields can also be substantially higher at the spot where a line turns at an angle, because of the way electromagnetic fields interact with each other.

Standards

The United States has no national standards for exposure to 60-hertz electric or magnetic fields, but some states have set standards for electric fields near rights-of-way. As of now, only two

Right-of-Way (RoW) Field Limit	
Montana	1kV/m at edge of RoW in residential areas
Minnesota	8 kV/m maximum in RoW
New Jersey	3 kV/m at edge of RoW
New York	1.6 kV/m at edge of RoW
North Dakota	9 kV/m maximum in RoW
Oregon	9 kV/m maximum in RoW
Florida	10 kV/m (for 500-kV lines) maximum in RoW
	8 kV/m (for 230-kV lines) maximum in RoW
	2 kV/m at edge of RoW for all new lines
	200 mG (for 500-kV single circuit) maximum edge of RoW
	250 mG (for 500-kV double circiut) maximum edge of RoW
	150 mG (for 230-kV lines) maximum edge of RoW

states have magnetic-field standards, and many say that they are so lenient as to be nearly meaningless. Limiting the electric field around a high-voltage line will also limit the magnetic field, however. But magnetic-field standards must be based on a line's current (which is what produces the magnetic field), not on the voltage. The right-of-way table shows the existing state standards for electric fields.

New York State has proposed interim magnetic-field standards for transmission lines and substations. But critics say that the proposals do nothing more than sanction the magnetic fields already existing and merely allow utility companies to say "we comply with the standards." Nevertheless, they are a step in the right direction—standards can always be tightened. The proposed limits are: (1) 150 mG at the edge of a 230-kV or smaller right-of-way, or at the edge of a new substation; (2) 200 mG maximum at the edge of a 500-kV right-of-way, or at the edge of a new substation serving a 500-kV line. In addition, New York has introduced a bill that would grant authority to the state's Department of Environmental Conservation to prohibit radar facilities or other sources of nonionizing radiation that are determined harmful to the public health or the environment.

Some states have tried to address the issue of schools near overhead power lines. Illinois introduced a bill that would require utility companies to move transmission lines at least 500 feet from schools and would set a 2-milligauss magnetic-field limit for residences. Oregon's Bonneville Power Administration has imposed a moratorium on playgrounds in rights-of-way. Some states have moratoriums on new lines altogether; others limit them to industrial zones. Many states now insist that utilities run new lines away from heavily populated areas and children's facilities.

California's Department of Education has established these setbacks for new schools near power lines: 100' from a 100–110-kV line easement; 150' from a 220–230-kV line easement; 250' from a 345-kV line easement; 350' from a 500-kV line easement. While such distances are a step in the right direction, many think they are still inadequate. A more thorough approach would be to set magnetic-field limits based on actual field measurements.

It is also important for a community to know whether a utility company plans to increase the amperage, or load—and hence the magnetic fields—of pre-existing lines. Utilities do not have to inform communities in advance, and states might consider legislation like the right-to-know laws now being applied to chemical companies. Some states require chemical companies to tell nearby communities what they are releasing into the environment and to give a warning twenty-four or forty-eight hours before each release.

Another approach that states have employed for setting utility-line standards comes from the nuclear industry. It is called ALARA, which stands for "as low as reasonably achievable," and is a guideline rather than a form of law. Although electric power is not really comparable to chemical or nuclear exposures, there is nothing wrong with borrowing concepts protective of the public from those industries.

High-Current Homes

Even homes and apartment complexes that are not close to power lines can be high-current dwellings because of certain common wiring configurations between the distribution line and the service drop to the home. Several studies have found increases in leukemia and brain tumors in children who live in such homes. High-current homes include:

- The first or second dwelling on the line from a pole-mounted step-down transformer (typically round and cylindrical, but sometimes open insulators on a platform).
- Dwellings within 15 meters (about 50 feet) of a thick three-phase primary distribution line; such lines are found at the top of wooden poles that run throughout a neighborhood, and segmented porcelain insulators usually connect them.
- Dwellings within 15 meters (about 50 feet) of six or more primary lines, also located at the top of the pole.
- Dwellings within 25 feet of a thin three-phrase primary distribution line.
- Dwellings less than 7.6 meters (about 25 feet) from three to five thin primaries, or high-tension lines with current capacities of 5 to 230 kilovolts.
- Dwellings less than 15 meters (about 50 feet) from 240-volt secondary lines that run directly from a pole-mounted step-down transformer to a home without losing power before passing the house.

These distances will seem small to someone living in a rural environment, but many city dwellers are often within 10 feet of primary lines. In some instances, for lack of space, a utility has even strung primaries directly on buildings. If you identify your home as a high-current one, be sure to take measurements, and try to rearrange the living space so that more time will be spent in the rooms with the lowest exposure.

The best prudent-avoidance advice is to keep exposures as close as possible to 1 milligauss in areas where the most time is spent. Readings above that—2 to 3 milligauss and up—have been associated with dose-related increases in certain childhood cancers. But remember that there is nothing inherently "health-ful" about the 1-milligauss level, only that bioeffects have been observed above that number, although the increases are small.

We may eventually find that 1 milligauss is too high, or even too low, or that slightly higher magnetic fields act to open crucial biological windows in cellular activity that are as yet imperfectly understood. Because of window effects, there is even the possibil-ity that we could be worse off by reducing certain higher expo-sures to specific lower ones. It is certainly a challenging area for researchers.

Studies

Hundreds of animal and human studies investigating the 60-hertz frequency have been done in the last few decades. Ap-proximately thirty studies have found adverse health effects for occupational EMF exposures (see Chapter 13). Those who live near high-tension lines or close to certain distribution lines have also experienced long-term, high-level exposures that correlate with many of these occupational observations.

The saga of positive versus negative studies (is there an ad-verse health correlation or not?) is a highly charged political is-sue, with accusations of bias from the power industry and counteraccusations of rigged studies from those outside the industry. Attacks on methodology by scientists reviewing each other's work are common. Interference from both the Reagan and the Bush administrations with the EPA's attempts to clarify the issue and warn the public is also documented. (For more information on "power-line politics," see anything written by Paul Brodeur. Dr. Robert Becker also explores the subject in *Cross Currents*, as does Ellen Sugarman in her *Warning: The Electricity Around You May Be Hazardous to Your Health.*)

Part of the problem with the 60-hertz frequencies (which fall in the extremely low frequency range usually defined as running from zero to 300 hertz) is that some mainstream scientists find it hard to believe that such low frequencies could be responsible for any biological damage. The wavelength of this frequency is about 3,000 miles long, and from the scientific standpoint, the only thing it could resonate with would be something resembling an earthworm just as long. Also, the actual energy content of 60 hertz is considered low (when compared with something like X-rays). But this frequency has turned out to have some interesting characteristics that probably act in synergy with naturally occurring DC magnetic fields from the earth.

Here are some of the more recent epidemiological studies of populations living near power lines:

- Dr. Stephen Perry, of Britain's National Health Service, observed that patients living near power lines had a high incidence of mental problems and a high suicide rate. Follow-up epidemiological studies with Dr. Robert Becker showed the same pattern. Also, studies by Dr. Charles Poole, of Cambridge, Massachusetts, found a strong link of nearly threefold more depression in people living near overhead transmission lines than in control subjects. These tend to verify occupational studies that found depression in electrical workers.

- Epidemiologist Dr. Nancy Wertheimer, at the University of Colorado, and physicist Edward Leeper discovered that 60-hertz magnetic fields of only 3-milligauss strengths were significantly related to the incidence of childhood cancers, especially leukemia and brain tumors. Wertheimer's studies were shocking—some say even to herself—and she called for review and replication. She and Leeper were the first to find such increases in relatively common residential exposures and at strengths considerably less than those common in households at the edge of some rights-of-way. Her studies correlated the childhood leukemia cases with utility-company wiring-code configurations (which indicate the amount of current that a line is capable of carrying). This correlation indicated that children living in "high-current" residences, which were the first or second on the line from a pole-mounted step-down transformer, were two to three times more likely to die of cancer than children in lower-current homes. Her methodology was criticized for not using actual field measurements taken in the homes (although Leeper did take many selective measurements to verify the wiring-code information), but her approach later turned out to be the most accurate one.

- Dr. David Savitz, of the University of North Carolina, repeated the Wertheimer-Leeper study (with vastly more resources at his command) and found the same results. It had been widely presumed that he would disprove her work, and so when he found the same increases it was as if a small bombshell had gone off. Savitz reported that 15 to 20 percent of childhood cancers appeared to be produced by exposures to 3-milligauss, 60-hertz magnetic fields—strengths common in some homes.

- Drs. Stephanie London and John Peters, at the University of Southern California, also supported the Wertheimer-Leeper and Savitz studies, finding a 115 percent increased risk of childhood leukemia associated with high-current wiring codes.

- Swedish researchers Dr. Anders Ahlbom and Maria Feychting, of the Institute of Environmental Medicine at the Karolinska Institute in Stockholm, found that children who lived near high-tension lines and were exposed to average magnetic fields of 3 milligauss or more had close to four times the expected rate of leukemia. The increases were incremental, and a clear dose-response relationship was found: those exposed to more than 1 milligauss had twice the expected rate; those exposed to 2 milligauss had nearly three times the rate; and those exposed to 3 milligauss had close to four times the incidence of leukemia of those exposed to only 1 milligauss. This study was painstakingly careful in using computer models and spot measurements as well as actual proximity to power lines. It also factored in air pollution and socioeconomic status, which did not alter the findings. The association with the incidence of adult leukemia was less strong. And, unlike other studies, this one did not find a link between power-line exposures and the incidence of childhood brain tumors.

- Dr. Jorgen Olsen and colleagues, at the Danish Cancer Registry in Copenhagen, found that children who lived near high-voltage facilities had a statistically significant fivefold increased risk of developing Hodgkin's lymphoma with average exposures of 1 milligauss or more. They also found a 40 percent increased risk of leukemia, brain tumors, and lymphomas in combination with average exposures of 1 milligauss or more, but the figures did not become statistically significant (at 5.6 times the expected rate) until the exposures reached 4 milligauss or higher. Above 4 milligauss, the leukemia and brain tumor risks were six times the expected rate.
- Finnish researcher Dr. Pia Verkasalo and colleagues, at the University of Helsinki, found similar patterns in children living near high-tension lines, but the study included relatively few cases, so the findings were not considered statistically significant. Nevertheless, the researchers thought they had found a two- to threefold relative risk of childhood cancers.

Numerous other studies focusing on the 60-hertz frequency have important implications for those with chronic power-line exposures. Here are some of them:

- Dr. Wendell Winters, of the University of Texas, investigated the effects of 60-hertz fields on human immune-system cells and also on human cancer cells in lab cultures. The cancer cells increased their growth rate within twenty-four hours by several hundred percent, after which the accelerated growth rate appeared to be permanent.
- Dr. Jerry Phillips, of the Cancer Research and Treatment Center in San Antonio, Texas, confirmed Dr. Winters's work and further found that human cancer cells permanently increased their growth rate by 1,600 percent and developed more malignant characteristics after being exposed to 60-hertz magnetic fields. Winters and Phillips both verified other work that had found EMFs to be cancer promoters in already malignant cells.
- Dr. Kurt Salzinger, of the Polytechnic University of Brooklyn, found that rats exposed to 60-hertz fields during fetal development and for the first few days after birth developed significant and unmistakable learning disabilities long after the time of the exposure.
- Dr. Frank Sulzman, of the State University of New York, found that monkeys exposed to 60-hertz fields had alterations in their biological cycles and activity levels that lasted for months after the exposure.
- Drs. Richard Seegal, Robert Dowman, and Jonathan Wolpaw, of the New York State Department of Health, found significantly depressed levels of certain metabolites of seratonin and dopamine (both are important neurotransmitters linked with motor activity, mood, and sleep, among other things) in monkeys exposed to 60-hertz fields. Only the dopamine returned to a normal level; the seratonin remained suppressed for several months after exposure. A lowered level of seratonin is suspected in depression and suicide incidence. This

research appears to verify Dr. Perry's and Dr. Becker's work on increased suicide rates in people who live near power lines. (It may also have correlations with air-ion research.) It was the first study to link 60-hertz fields with alterations in neurologic function.

• Dr. Richard Phillips, while at the Battelle Pacific Northwest Laboratories, conducted studies first on mini-pigs, and later on rats, exposed to 60-hertz fields specifically designed to simulate power-line fields. Nearly the entire group of exposed mini-pigs died in an epidemic that unexpectedly swept through the group, while far fewer of the nonexposed control group died. The implication was that the exposed population had a lower immunity. The subsequent follow-up study on rats, although called "inconclusive" in the official report, was later said by Phillips to have found a suppression of nighttime pineal-gland melatonin production (in rats exposed to 60-hertz fields for three weeks); a significant reduction in blood levels of testosterone (in male rats exposed for three months); alterations in the neuromuscular system of animals exposed for thirty days; and increases in birth defects in both rats and mini-pigs that were chronically exposed for two generations.

• Drs. Robert Becker and Andrew Marino continuously exposed rats to 60-hertz electric fields for three generations and measured the results against a control sample. They found increases in mortality and lower birth weight in the exposed animals.

• Dr. José Delgado, at the Centro Ramon y Cajal in Madrid, found major developmental defects in chick embryos exposed to 10-, 100-, and 1,000-hertz magnetic fields at extremely low field strengths. Most malformations occurred at around 100 hertz at strengths as low as 1 milligauss. (The work was later continued by Dr. Jocelyne Leal, who found the same patterns.) The U.S. Navy commissioned studies from six different laboratories to replicate the work; five of the six labs found that very low level, very low frequency, pulsed magnetic

fields contributed to increased abnormalities in early embryonic chick growth.
- Dr. Reba Goodman, of Columbia University, in continuing
work over a number of years, has found chromosomal defects
in both human and insect cell cultures exposed to ELF frequencies. The relationships are complex and depend on different frequencies as well as different cell types.

What many of these studies tell us is that two different
anatomical systems are primarily affected by extremely low
frequency, 60-hertz fields—the brain and its neurohormo-
nal agents, and growing tissue cells, including fetal cells, fast-
growing cells in children, and the malignant cells of cancer.

Wiring Configurations and Mitigation

As said before, mitigation techniques for power-line exposure
are mostly up to the utilities. Their options include burying the
lines underground; creating larger rights-of-way; and canceling
fields by rephasing the current on the lines, redesigning the way
wires are hung, and adding new wires to existing configurations.
All these options are being actively investigated by utility companies, but there is still much resistance to change.

Burying. The burying of large transmission lines underground is the most expensive option, although underground distribution lines have been used in cities all along and are widely
found throughout Europe. The cost of burying high-tension cable
over a long distance can run up to $5 million a mile. The cable is
encased in one of three different types of pipe—gas-insulated,
oil-insulated, or pressurized. The electric fields from underground cable are essentially zero, and magnetic fields are tiny.
The ground above the cable is suitable for livestock grazing and
some kinds of agriculture, but buildings or trees with deep roots
are not allowed. Underground cable requires aboveground cool-
ing equipment, which produces some noise (from fans) and

EMFs of its own. Underground cables are more weather resistant than aboveground lines. When something goes wrong, however, locating the source of the problem can be more difficult, which means that power outages will be longer (although far less frequent than outages for our aboveground system now). Repairs are far more difficult and can involve extensive digging—next to impossible if the ground is frozen. Also, depending on the pipe used, gas or oil leaks can occur if the earth shifts in a geologically unstable area. Except over relatively short distances, the undergrounding of transmission lines is considered impractical by utility companies, although it remains the popular choice of consumers. The high costs that utility companies cite would probably decrease if the companies factored in the cost of repairing overhead lines downed by storms, which can take days and sometimes weeks to repair. There is increasing pressure on utility companies to bury primary and secondary distribution lines, too.

Larger Rights-of-Way. Many states now require that utility companies have more land around power lines, especially when new lines are sited. Some homeowners have been successful in getting utilities to purchase their property at a fair market value if the property cannot be sold within a certain period of time. Many of these arrangements have had gag orders imposed on them by utility attorneys, however, so much remains unknown about specific details. Depending on what state you live in, a utility company may be more or less willing to strike such a deal—and it will take an attorney to negotiate it for you.

Wiring and Rephasing. By far the least expensive mitigation approach is to rephase the current on three-phase circuits or hang wires in a way that will allow the fields to cancel each other out. The strength of magnetic fields depends on the power line's load of current, its height, and the way the wires are distanced from each other. The higher the current and the farther apart the wires are positioned, the higher the magnetic fields will be.

In discussing mitigation techniques through wiring rearrangements, it is important to compare power lines carrying the same current. Transmission lines capable of producing the highest magnetic fields are single-configuration three-phase circuits, which take two forms: three vertically stacked lines on one side of a tower, or three wires on a single tower in a vertical arrangement. High-EMF distribution lines are the classic T-shaped poles running along a roadway that we are all accustomed to. These are single circuit. Double-circuit lines are three vertically stacked wires on both sides of a tower or in a two-tiered horizontal arrangement. If the current on the lines is optimally phased and balanced, double-circuit lines have the greatest potential for magnetic-field reduction. Single-circuit lines can also be altered to reduce fields.

How the wires are spaced, no matter what the current load, is the most important factor in magnetic-field reduction. The best tower style for such mitigation is called the delta—named after the fourth letter in the Greek alphabet that looks like a triangle with three equal sides. Deltas can be either vertical or horizontal; either one will bring the wires closer together for a cancellation effect. Some vertical deltas, when phased properly in a double circuit, have the potential for very low magnetic fields—even lower than some common domestic backgrounds. (See the illustration called "Wiring Configurations" on the following page.)

As for the current phases, with no adjustments in the way the wires hang it is still possible to reduce magnetic fields through what is called reverse phasing. Here's how phasing works: fields that fluctuate together at the same strength, frequency, and direction can add to each other. This is called matched phasing, and it creates the highest magnetic fields. Fields that are exactly out of phase with each other, meaning that each one reaches its greatest strength at a different time, achieve a substantial amount of cancellation. This is called reverse phasing, and it is being used by many utility companies. (The illustration titled "Rephasing" shows how this works.) One of its advantages, besides its low cost in comparison with other mitigation techniques, is that it can be done right at a substation, with no rewiring needed. Reverse phasing works for both horizontal and vertical configurations.

Other mitigation techniques available to utility companies include: placing high-voltage and low-voltage lines close together, which will help them cancel each other out; placing lines closer together on towers (compaction or bundling) by using V-string suspensions; grouping new high-tension lines along existing corridors rather than putting them in new areas; or altering the amount of current carried on some lines. Additional concepts are still in the research phase, such as twisting some wiring configurations and running additional wires to help cancel fields.

Mitigation for single-phase distribution lines that serve very

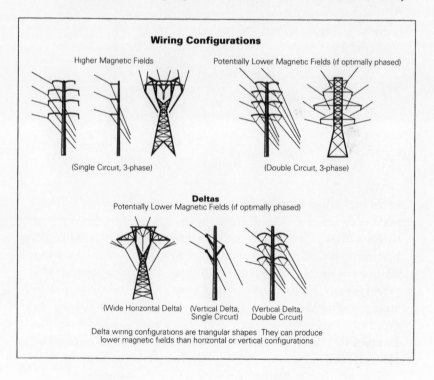

Wiring Configurations

Higher Magnetic Fields

Potentially Lower Magnetic Fields (if optimally phased)

(Single Circuit, 3-phase)

(Double Circuit, 3-phase)

Deltas
Potentially Lower Magnetic Fields (if optimally phased)

(Wide Horizontal Delta)

(Vertical Delta, Single Circuit)

(Vertical Delta, Double Circuit)

Delta wiring configurations are triangular shapes. They can produce lower magnetic fields than horizontal or vertical configurations

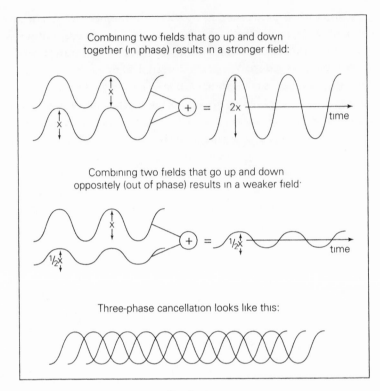

Rephasing

small areas (like a single street or neighborhood) is more difficult and challenging. Currents typically are not balanced or phased, so rephasing may not produce the hoped-for results. Also, the matter of grounding comes into play. (See Chapter 12.) A current improperly grounded between buildings can follow unintended pathways on domestic and neighborhood water pipes, for instance, rather than returning to the poles on the utility's neutral ground wire as it's supposed to. Mitigation for single-phase lines involves the better balancing of current loads and return currents. If you get high readings in your home, call your utility company or an independent consultant to help determine the cause.

All these approaches have their advantages and disadvantages. It is up to the utility company to decide what technique is best suited for each situation. But if you are in a dispute with a utility company and it offers to rephase the lines as a way to reduce magnetic fields, take the company up on the offer.

At the moment, most utility companies are not highly motivated to reduce electromagnetic fields voluntarily. They reason that the adverse effects shown in most studies are small and that common domestic fields are often higher than those created by some power lines—especially when consumers "intentionally" increase their exposures by using such items as electric blankets, or do not follow through on "unintentional" exposures such as building-code wiring violations.

The ultimate long-term solution to cut down on transmission lines is to have many more generating sources in local regions so that current would not have to be moved over such long distances. Generation sources could be developed to best suit a region's resources—solar generation in the southwest and hydroelectric in the northeast, for instance.

Chapter 15

THE BLINKING HORIZON . . .
RADIO TOWERS AND MICROWAVE
DISHES

R ADIO TOWERS have begun to capture the consumer concern
along with power lines. We see more and more towers with
several dishes mounted on them blinking on the horizon,
turning our formerly pristine ridge lines into occupied zones of
industrial blight. They perch atop our hospitals, municipal buildings, apartment houses, police and fire stations. They seem to pop
up overnight on school property or appear in vacant fields near
our homes. They dot the highways like bizarre outcroppings. We
see single towers, small groups of varying heights and styles, and
whole antenna farms. Who hasn't come around a bend on a country road and been abruptly confronted by a tower of some kind
where none was expected?

Their shapes and designs are myriad. Sometimes they look
like large books and are mounted on the side of a building. Others
are tall and strange-looking and have been invited into a neighbor's backyard. Some are small and straight and are mounted on
relatively low structures. Others look disproportionately tiny and
anticlimactic when mounted on a tower that would put a small
skyscraper to shame. Some are shaped like round dishes. Others are mounted on platforms that have lots of small antennas
reaching skyward, like an electronic fright wig. Some have

even been woven into deceptive designs not unlike Victorian metal grillwork, to make them more attractive but also to camouflage them so that concerned consumers won't recognize what they are.

Few of us recognize what we're looking at. Just what are all these antennas in our midst? And what are they doing to us and to our environment?

Transmitters

We have all become accustomed to electronic communication, be it from radio, TV, on-line computer services, or wireless devices. We rely on traffic reports at rush hour and weather forecasts on TV, especially when we can see something as dramatic as a hurricane being tracked on radar. We also presume that when we fly, the airplane won't crash into other airborne planes and that we will arrive at our intended destination, not somewhere else. We also assume that our military will be able to detect any aggression by a foreign country (like a missile fired at the White House) in time to stop it. On the local level, when a police, fire, or medical emergency occurs, we assume that a request for help will be met with a response.

What all these situations (and many others) have in common is that they are made possible through radio/microwave frequency transmission. While the power and frequencies of these technologies differ, in some ways they are similar. All generally fall within the radio-frequency range of the electromagnetic spectrum, and they have proliferated at an alarming rate within the last two decades—without any clear understanding of their bioeffects and without adequate federal standards for safety, despite disturbing research that goes back to the 1940s. We simply don't know what a safe level of RF radiation is, yet we continue to increase our exposure to it every day.

How Radio and TV Transmission Works

Radio waves are produced when an electrical signal is fed to the mast or antenna of a transmitter. This electrical signal (when oscillated) causes the electrons in the metal atoms of the mast or antenna to change their energy level and emit radio waves. Radio and TV transmitters broadcast waves that are modulated, that is, the original sound or light signal is superimposed on the radio wave (the carrier wave) in a way that carries it. The receiver at the other end, like a radio or TV, removes the original signal from the carrier wave, and that is what we hear or see.

In the radio studio, a microphone responds to the sound waves of a human voice by producing an electrical signal that changes in voltage at the same frequency. This modulated signal is then amplified, and the signal is sent to the transmitter. Such signals used to be sent to the transmitter (typically located at a distant site) by electrical lines, but today they are sent by a station transmitting link (STL), which is often a microwave transmitter located right at the station.

The frequency of the carrier signal is constant and much greater in strength than the frequency range of the sounds being carried. The carrier frequency is an assigned place (set by the FCC) on the spectrum. That is the number that the station reads on the air and the number that is on your dial.

There are two kinds of modulation: amplitude modulation (AM), in which the amplitude (energy level) of the carrier wave varies at the same rate as the changing voltage in the sound signal but the carrier frequency remains the same; and frequency modulation (FM), in which the carrier waves themselves vary with the voltage level of the sound but the amplitude remains the same. The illustration shows how both AM and FM modulation work.

AM waves tend to hug the ground and follow the curvature of the earth. Therefore AM antennas are often located in valleys and are more suitable for hilly or rural areas. FM signals, on the other

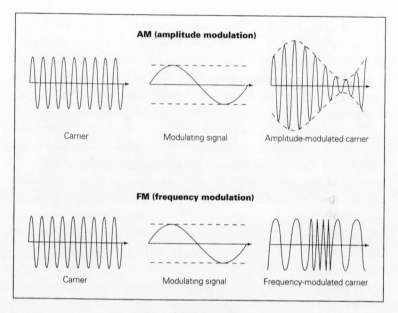

AM and FM Modulation

hand, as well as TV and microwaves, travel in line-of-sight paths. Therefore their antennas are placed on the highest ground available, so they can blanket an area. That is why we see a number of towers clustered together on a hilltop outside a city or on top of the tallest buildings. (An exception is cellular-phone towers, which often line the highways.)

In general, an antenna tower can be thought of as the jumping-off point for modulated signals. Radio towers can broadcast several carrier waves at different frequencies, with each carrying a different sound signal, but each radio station or TV channel broadcasting from the transmitter has a different frequency. Both AM and FM radio can carry either analog (steady) or digital (on/off) signals. With the growth of digital technology, it is now the digital signals—which simulate pulsed waves—that pose an additional health concern. And modulation itself, whether AM or FM, is also a serious concern, because carrier waves can effectively infuse the atmosphere with lower frequencies, which may be bioactive to different species (including humans) even at low power intensities. In addition, the FCC does not measure the modulated signal, only the carrier signal—on the rare occasions when any measurements are taken.

The purpose of the tower is to support the transmitter (the tower itself is not a transmitter). The tower might be a single symmetrical tower reaching hundreds of feet high (painted red and white, with flashing lights on top, if it exceeds 150 feet), on which are mounted TV and FM transmitters at the highest point, with perhaps several microwave dishes at lower mountings; or it can be a series of vertical spires positioned so as to interfere with each

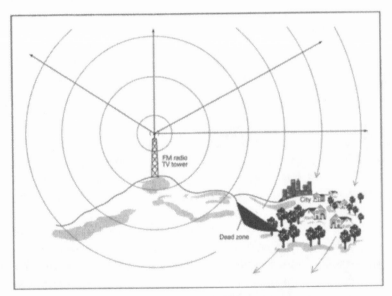

Radiation patterns from an FM and TV transmitter spread out in straight paths in all directions. Receivers must have clear "line of sight" to transmitters for signals to be received. "Dead zones" occur when signals are blocked by mountains, hillsides, some buildings, or other obstructions.

ADAPTED FROM *CROSS CURRENTS*, BY DR. ROBERT BECKER.

other in a way that changes the radiation pattern. These latter arrangements, called directional antennas, are intended to direct transmissions away from such things as water tanks, buildings, or land areas that a station wants to avoid when it would interfere with another station's territory. Sometimes signals will be relayed from one tower to another, which serves to increase the broadcast reach of a particular station. But most transmitters radiate in a 360-degree pattern, even the line-of-sight towers.

Antenna radiation is usually in looplike patterns of varying lengths. But many towers also create side lobes or radiation leakage, in ways that engineers often do not foresee. These side lobes can contribute a substantial amount of radiation to anyone living or working near a transmitter.

RF transmission signals are present in the environment all the time. Just turning off your radio or TV does not mean that the exposure disappears. If you get good radio and TV reception without a cable hookup, you live in a high-exposure environment.

What Radio-Frequency Radiation Is

Radio-frequency (RF) radiation falls between about 3 kilohertz and 300 gigahertz on the electromagnetic spectrum. (See Chapter 3 for the complete electromagnetic spectrum.) This is a broad area on the spectrum which includes many frequencies and wavelengths.

All radio waves radiate into space at the speed of light (186,000 miles per second), and each wave is associated with a wavelength and a frequency. Wavelength refers to the distance that the wave travels through space in a single cycle between one peak or valley of the radio wave and the next. Frequency indicates the number of waves in one second. The following illustrations show the various radio frequencies and their uses.

A simple mathematical formula is often used: frequency times wavelength equals the speed of light. Since the speed of

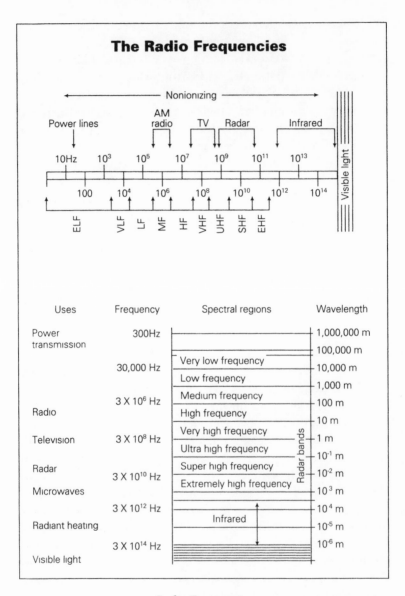

Radio Frequencies

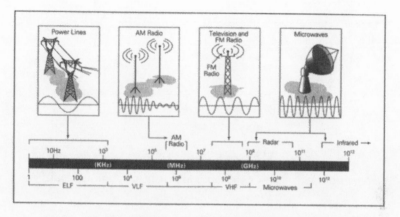

*Where different radio-frequency technologies are located
on the electromagnetic spectrum.*

light is a fixed number, the frequencies and wavelengths have an inverse relationship. The higher the frequency, the shorter the wavelength; the lower the frequency, the longer the wavelength. The higher up on the spectrum one gets, the shorter the waves become and the more power they inherently have.

Since the radio-frequency range is such a broad one, the wavelengths vary considerably. There are long waves, medium waves, and short waves, and further categories called extremely low frequency (ELF), very low frequency (VLF), low frequency (LF), medium frequency (MF), high frequency (HF), very high frequency (VHF), ultra high frequency (UHF), super high frequency (SHF), and extremely high frequency (EHF). The RF bands are usually defined as 300 megahertz to 300 gigahertz, although some lower bands are often included. Microwaves occupy the ultra, super, and extremely high frequency bands; radar is part of the microwave band. FM radio uses VHF; TV uses VHF and UHF; cellular phones and satellite broadcasting use the UHF-to-microwave frequencies. The following boxed list shows the uses common to each of the frequency bands:

Frequencies

Very low (VLF)	Time signals, very long range military communications
Low (LF)	Air and marine navigation
Medium (MF)	AM radio, air and marine communications, SOS signals, ham radio
High (HF)	International shortwave, long-range military communications, ham radio, CB radio
Very high (VHF)	FM radio, VHF TV, police and taxi radios, air navigation, military satellites
Ultra high (UHF)	UHF TV, international communications, police and taxi radios, microwave ovens, radar, weather satellites, cellular phones and towers
Super high (SH)	Commercial satellites, microwave relay, air navigation, radar
Extremely high (EHF)	Future uses like personal communications systems, short-range military communications

Radio waves are often used in relationship with the ionosphere, with lower ranges of around 50 miles above the earth and upper ranges of 300 miles. FM and VHF-TV waves penetrate the ionosphere; medium waves are absorbed by it; short waves are reflected repeatedly between the ionosphere and the earth. The illustration on the following page shows these interactions with the ionosphere.

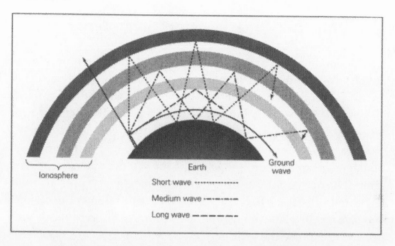

How Different Radio Waves Interact with the Ionosphere

Radio waves are used in a number of ways to accomplish different ends. Long ground waves have a range of thousands of miles and will curve around the earth's surface. VHF medium waves travel short distances and bounce off the ground or large objects. Medium "sky" waves can be reflected off the ionosphere to come back to earth in a completely different place. Short waves are multiple reflections that bounce between the ionosphere and the earth in different locations, allowing for worldwide communication in that frequency. Radar bounces off objects and returns to the transmitting source, creating an image on a screen; thus, some radar transmitters produce a broad spray of microwaves over a huge area.

Radio waves are not always as well behaved as broadcasters would like. During a solar storm the ionosphere may reflect waves thousands of miles off course, and listeners will pick up stations or signals from halfway around the world. There is also some indication that the earth's magnetosphere (above the ionosphere) may amplify some of these ever-increasing RF sources in a way that is changing our weather patterns. (See Chapter 5 for more details.)

Another Quick Review of Some Terms

Consumers who have educated themselves about power lines are often struck by what seems to be a completely different technical language when they begin to look into the radio frequencies—as if it's a different science. Power-line frequencies are measured in milligauss, gauss, and teslas, whereas radio frequencies are measured in microwatts, milliwatts, and distance relationships in square centimeters. But the science (physics) is the same. The reason for the difference in terms is partly professional territoriality, but also because waves behave differently the higher one gets on the electromagnetic spectrum, so different terms are needed to capture those principles.

Frequency is measured in hertz:

1 hertz (Hz)	= 1 cycle per second
1 kilohertz (kHz)	= 1,000 hertz
1 megahertz (MHz)	= 1 million hertz
1 gigahertz (GHz)	= 1 billion hertz

Radio-frequency radiation has both an electric and a magnetic component. Volts per meter (V/m) is the unit of measurement for the electric component, amperes per meter (A/m) for the magnetic component. Combinations of the two that exist in space are called electromagnetic fields, and field strength is the measurement at a particular location. Power density is the term used for measurements of the far-field zone of radiation—a point far enough away from a transmitter so that near-field (closer to the transmitter) interactions don't apply, and at which the electric and magnetic components form plane waves that radiate into space.

The environment around antenna sites (called "near" fields)

can be an electromagnetically complex one, to say the least. Radio waves couple with each other as well as with other frequencies, and in some circumstances (like the presence of metallic objects) can create what are called standing waves, meaning that they do not radiate into space, but remain within a set area at high intensities. Couplings can add to or subtract from each other, and complex fields of high power intensities can and regularly do interfere with nearby radios and TVs—and undoubtedly with a person's innate biological fields as well. There is some indication that certain FM frequencies may create standing waves in the brain that the body cannot dissipate. It is also one of the concerns about cellular phones that use microwave frequencies.

Power density is measured in terms of units of power in a set area, typically in milliwatts per square centimeter (mW/cm^2). Sometimes microwatts per square centimeter (μM/cm^2) are used. (Milliwatts are 1,000 times more powerful than microwatts.)

The power-density measurement is often linked to its ability to heat human tissue. RF/MW energy is absorbed by the human anatomy—described by the term specific absorption rate (SAR). Standards based on tissue-heating reactions (thermal effects) set limits on power-density exposure based on SARs averaged over a certain length of time. Standards today range around 0.4 watts per kilogram (0.4 W/kg) over a continuous six-minute period, which is an expression of the energy absorbed by the body during that time. There are SARs for specific areas of the body, such as the head, and there are whole body absorption rates, too.

One criticism of such a standard is that it allows for the power intensity to exceed that limit if it is for less than six minutes. But no one really checks the time period—and it is the higher intensities for short durations, which simulate pulsed exposures, that are of the greatest concern. Anyone living near a cellular transmitter, for instance, would experience short, intense exposures all the time.

A Research Vacuum

Radio-frequency radiation is likely to prove to be a more hazardous environmental and health risk than the power-line frequencies for several reasons.

There is far more research on the power-line frequencies than on RF/MW, and many observers of the politics of research say that this gap looks suspicious, that the long arm of the U.S. military has contributed to it. The military traditionally has not been involved with the power companies in the same way as it has with radar and communications technologies. Consequently, the air of secrecy—if not outright sabotage—never took deep root in the utility companies the way it has with communications. The utility companies seem to want clear bioeffects answers about EMFs, whereas the military has historically appeared to prefer an absence of biodata, at least in civilian hands. The entire course of communications development has been allowed to burgeon in this research vacuum. The utilities also appear to be less calloused toward the civilian population than the military has sometimes been. (It is hard to imagine the utility consortiums willfully increasing the electrical current in an area just to see what happens, the way the military has experimented on some localities with nuclear fallout.)

In the past, the paltry RF research that was done concentrated on the microwave bands. But the lower RF frequencies are finally beginning to get some attention, now that the cold war is over. Environmental measurements of the ambient background have shown that RF levels, especially in the FM band, are the ones of greatest magnitude today. Every new transmitter that goes on-line adds to this background, and consumer pressure is becoming a big factor.

Most, although not all, RF/MW laboratory research has been done on unmodulated carrier waves, and the bioeffects reported have been for fields of high enough intensity to cause tissue heating. But when carrier waves are modulated with other frequencies, a whole range of biological observations becomes possible, perhaps because we are more attuned to them in nature, but also because it brings in all of the other research done at the lower power-line frequencies. This is an important point because it now links a larger store of scientific information with previous, seemingly unrelated research.

This would apply to research on the environment and other species as well. We have not made much attempt to identify the resonant frequencies of other species, but there is the possibility that we are triggering critical parameters in some of them. For instance, frogs are disappearing all over the globe. Research in the 1920s discovered that frogs' eggs were affected by the 20-hertz frequency. Is there a resonant connection with a new modulated carrier signal in that band? The former Soviet Union operated a huge transmitter (nicknamed Woodpecker) that was modulated at 10 hertz in various pulsed combinations of 6 + 4, 7 + 3, and so on. Is there now some other combination at around 20 hertz initiated by someone else? And what about other frequencies? More recent research conducted in the 1970s by Dr. Allen Frey found that microwaves could alter the heart rhythm of frogs—including stopping it altogether—when the pulse was synchronized with the heartbeat. Numerous studies have found changes in the blood-brain barrier in test animals exposed to microwaves. Yet over one hundred thousand new cellular phone towers alone are planned in America—all broadcasting in the microwave frequencies.

That we may be triggering such critical parameters in other species with these frequencies is an interesting line of thought that perhaps a curious researcher will want to follow.

The Bioeffects of Radio Frequencies

Only a handful of epidemiological studies have been made of people living near transmission towers, and none of radio or TV broadcast technicians. Despite this research vacuum, many things are known about RF/MW exposures, from which we can make some reasonable extrapolations. The picture is a disturbing one. We have already discussed the early microwave research, in Chapter 2. Here is the more recent perspective.

The human anatomy can absorb significant amounts of energy from environmental RF exposures. The RF bands generally fall between 3 kilohertz and 300 gigahertz, and humans absorb radiation best between 30 and 100 megahertz, with the maximum absorption for the theoretical average person at around 77 to 87 megahertz. This is the FM radio band. (Women and children absorb radiation differently than this average model, but their anatomical differences are not usually taken into consideration.)

The amount of radiation absorbed will depend on proximity and orientation to the RF source, the person's shape and size, any couplings with other conductive objects, and the frequency and power density itself. As said previously, these can be complex variables in designing research models. Other variables include the area of the body being exposed, the water content of the specific tissue, the presence of sodium, lithium, or iron in that tissue, whether the tissue is fast-replicating or glandular tissue, and many other factors.

It has been known all along that RF radiation excites the water molecules in human tissue and causes heating, and most attempts at setting standards have been based solely on that aspect. But the nonthermal effects are the more important and biologically fascinating ones.

In humans, EMFs in various frequencies have been found to adversely affect calcium binding at the cell surface, DNA synthesis, and cell division; to alter circadian rhythms, affect or alter some important enzyme activities, and affect specific glands like the pineal and the hypothalamus area of the brain, as well as the production of certain neurotransmitters, like seratonin and dopamine; to increase the permeability of the blood-brain barrier; to create artificial stress responses; to overstimulate the immune system initially, then suppress it and decrease T-lymphocyte production; and to promote malignant tumor growth with particular concentrations in the central nervous system, in the blood and skeletal systems, and in glandular tissue. The eyes, the brain, and the testes seem to be especially prone to abnormal effects from the RF frequencies. The eye serves to amplify some RF/MW frequencies, which is probably why increases in posterior cataracts have been observed in some microwave workers. (Microwaves are also known to increase drug sensitivity in people taking glaucoma medication.) The testes are very close to the body's surface, which is probably why increases in testicular cancer have been reported in law-enforcement officers who have rested functioning radar guns in their laps. In addition, it appears that the human anatomy has specific windows of sensitivity at which certain bioeffects have been repeatedly observed, but not at other frequencies.

All these observations have been explored in detail in other chapters, as well as unique applications to women and children. Some studies, however, are worth specific mention here because of their implications for the long-term, low-level exposures that those living or working near transmitters experience. (A comprehensive report of studies on occupational exposure, which pertain to the RF/MW frequencies, is in Chapter 13. Many of these studies apply to civilian exposures, too.)

The Soviet Union did some interesting RF/MW research on behavioral aberrations that is unparalleled in the United States. It has been known for many years that low-intensity EMFs produce adverse effects on the autonomic and central nervous systems of humans and animals in strengths far too low to cause tissue heating. For years U.S. researchers dismissed much of the Soviet research, partly for political reasons but also because they could not replicate many of the studies because the Soviets (for security reasons of their own) left out important details. With the end of the cold war, some of these gaps have been filled and American researchers have been able to replicate some Soviet work.

Radio-wave sickness is the term the Soviet researchers used

to describe a clinical syndrome in those occupationally exposed to EMFs, particularly RFs/MWs. It included functional disturbances of the central nervous system such as headaches, increased susceptibility to fatigue, increased irritability, dizziness, sleepiness, sweating, concentration difficulties, memory loss, depression, emotional instability, mild limb tremors, cardiac arrhythmias, increases in blood pressure, and appetite loss. Thyroid enlargement, benign adrenal-gland tumors, and rashes were also observed. Less common but also reported were hallucinations, insomnia, fainting, and internal organ or intestinal difficulties. Also, auditory channels were stimulated when the head was exposed to low-power, pulse-modulated RF.

Of additional interest with regard to long-term, low-level exposures was research done with Polish career military personnel in the late 1980s by Stanislaw Szmigielski and co-workers, at the Center for Radiobiology and Radioprotection in Warsaw. The military personnel, whose major exposures were from the radar/microwave frequencies but with some 50-hertz exposures also involved, were found to have a six times higher cancer incidence than nonmilitary test subjects. Most of the malignancies were lymphomas and leukemias. Szmigielski is also known for work on EMFs and immune suppression that suggested a biphasic reaction of initial stimulation followed by overall suppression.

This work was similar to rat studies done by Dr. Arthur Guy, at the University of Washington, which found increases in cancer incidence as well as suspected immunological irregularities in exposed test animals.

In 1990, Dr. Vera Garaj-Vrhovac and co-workers, at the University of Zagreb in Croatia, found abnormalities in blood lymphocyte chromosomes in Yugoslavian microwave workers. The workers' exposures had been over periods of eight to twenty-five years, with intensities ranging from 10 to 50 $\mu W/cm^2$—approximately 1 percent of that allowed by the FCC. The same researchers produced similar chromosome abnormalities in laboratory cultures of mammalian cells at levels of only one-twentieth of the limit currently allowed by the FCC. The length of time these workers were exposed brought up the point that long-term, low-level effects were cumulative and would affect anyone living near a RF/MW transmitter, such as a radar, radio, TV, or cellular-phone tower.

The same issue of low-level, long-term exposure came up in the 1987 work of Terry Thomas and co-workers, at the National Cancer Institute, in their study of electrical and electronics RF/MW workers and an increased incidence of malignant brain tumors, especially astrocytomas. For those with exposures extending over twenty years, the risk was ten times higher than in the control group, especially if they were also exposed to soldering fumes and solvents. This suggested a joint action of chemical factors and EMFs, as discussed in other chapters. (Other American research has turned up both accelerated growth rates and suppression of brain-tumor cells at different RF/MW frequencies, and increased permeability of the blood-brain barrier, which normally prevents toxins from entering the brain—all at very low powers. Some tumor cells continued to proliferate at abnormally high rates—for as long as five days after only a two-hour exposure. There appeared to be frequency windows for both positive and negative effects, which indicates that some frequencies may lend themselves to therapeutic use.)

One of the more haunting American studies by far—one that has gotten very little attention—was reported by William Bise in 1978 in an article entitled "Low Power Radio-Frequency and Microwave Effects on Human Electroencephalogram and Behavior," in *Physiological Chemistry and Physics*. This study found that ten very carefully selected human volunteers experienced temporary changes in brain waves and behavior when subjected to different frequencies at different times of days in both shielded and non-shielded environments.

The year-long study (between July 1975 and June 1976) in-

cluded men and women ranging in age between eighteen and
forty-eight, from diverse occupational backgrounds. Three had
been occupationally exposed to very high frequency and microwave radiation; the other seven had not. (The group included
two health physicists, a housewife, one hospital intensive-care-
equipment expert, two secretaries, an electronics engineer, one
computer programmer, and two others.) Tests were conducted
with both modulated waves and continuous waves.

Changes were seen in brain-wave patterns from continuous-
wave radiation ranging from 0.1 megahertz to 960 megahertz
(which includes the cellular-phone frequencies), as well as from
pulse-modulated waves between 8.5 and 9.6 gigahertz. The most
striking effects were noted at between 130 and 960 megahertz
with continuous-wave exposures. Of particular significance was
that the power intensities that induced the most profound effects—10 pW/cm^2 (picowatts) and 10 fW/cm^2 (femtowatts)—
were far below the current FCC standards, and are typical of the
ambient background in many urban environments.

Alterations in brain waves occurred in both male and female
volunteers, with desynchronized alpha waves that were 15 to 25
percent higher than normal in some frequencies but were diminished in others. Slow brain-wave patterns also appeared at different amplitudes. Four of the male test subjects experienced
short-term memory impairment followed by an inability to concentrate and irritability. Three female test subjects experienced
irritation and severe anxiety. Using other parameters during the
experiment (near midnight with X-band pulse-modulated power
levels of 10 to 12 W/cm^2), three males experienced severe frontal
headaches as well as heart irregularities previously unseen
throughout the testing. The same men felt mentally and physically unable to work the following day. Within forty-eight hours,
the effects had reversed.

Other changes were noted throughout the experiment, depending on the time of day, whether the test area was shielded

from other potential EMF interference, and what frequencies and amplitudes were used.

It was a complex study that raises many questions and appears to verify the observations of other researchers, particularly the Soviets. The ethics of using human test subjects comes up, of course, but the intensities involved were so low as to be indistinguishable from the ambient background near most radio and cellular phone towers and were substantially below some common urban levels.

Typical RF Exposures

Those who live nearest to transmission towers have high-intensity exposures for twenty-four hours a day, varying according to what is being broadcast. Radio, TV, and microwave transmitters are widely scattered throughout most urban and suburban areas today, meaning that most of us are exposed to various degrees of RF/MW radiation.

In 1986, the EPA conducted a survey of fifteen large cities to determine just what the RF/MW exposure of the general population was. (The survey represented about 20 percent of the overall population.) The researchers used population-weighted average power levels in the urban areas and estimated the residential median exposure at 0.005 microwatts per square centimeter (0.005 μW/cm^2) for the FM, radio, and TV broadcast frequencies, and at .019 μW/cm^2 for the AM broadcast frequencies.

This sounded like good news because it meant that, of those surveyed, 99 percent were not exposed to levels above that suggested in the Soviet Union, which was the most stringent in the world at 1 μW/cm^2. The U.S.'s ANSI standard is 100 times more lenient, at 1 milliwatt per square centimeter (1 mW/cm^2), and could become more lenient still for some exposures. The remaining 1 percent in the EPA survey, however, were found to be regularly exposed to levels above 1 μW/cm^2, due to close proximity to a generating source of one kind or another. Also, exposures due to nearby sources in readily accessible areas (rooftops, adjacent buildings, and so on) may have reached maximum levels of between 1 and 10 mW/cm^2 or higher. The EPA reported that the number of people exposed to these levels was not known.

The study did not measure exposures at heights of more than six meters above the ground, and no attempt was made to measure standing waves or other near-field phenomena. In addition, the study was conducted before the advent of cellular-phone transmitters, which are mounted on many urban buildings today. The ambient backgrounds could have become significantly higher in the decade since this study was conducted, especially in high-rise buildings.

One of the more interesting findings of the survey was the fact that the FM radio frequencies were most responsible for the overall exposures, registering the highest measurements as the dominant waves at ground level. Keep in mind that humans are resonant with the FM band.

Those living or working near transmitters have higher exposures than those found in the EPA survey. By the estimate of Dr. Andrew Marino, in an assessment of possible exposures to residents near a cellular-phone tower in one California community, annual exposures could be between 1,350 and 18,000 times greater than the EPA's typical city exposure as found in its survey (the huge range was due to best-case, worst-case estimates by the cellular-phone companies). These exposures would be comparable to the occupational exposures already mentioned—levels at which serious health damage was observed.

Very few epidemiological studies have been made of people living near transmission facilities, but those that exist have found increases in adverse health effects.

- *Honolulu:* In 1984, the EPA measured RF levels at twenty-one sites near broadcasting towers in Honolulu and found that at two sites the levels exceeded the ANSI standard. Nearby residents had complained of RF-associated shocks and burns, as well as interference with electronic equipment. The EPA, using information from the Hawaii Tumor Registry and after adjusting for age, sex, and race (but not for smoking, diet, or occupational history), found that the incidence for all cancers in both men and women was significantly higher in those living near broadcasting towers than in those not living near such facilities.

Also in Hawaii, a cluster of childhood leukemia cases was identified in the mid-1980s near a U.S. Navy submarine communications facility called Lualualei, on the island of Oahu. The Hawaii Department of Health and the EPA found a statistically significant doubling of the incidence of leukemia in children who lived closer to the facility.

- *Thurso, Scotland:* A similar statistically significant cluster of childhood leukemia cases was identified in a north-coast Scottish town near a U.S. Navy communications facility.
- *Birmingham, England:* A cluster of leukemia and lymphoma cancers was identified in residents living near a 750-foot BBC radio and TV tower that had four TV and seven radio transmitters mounted on it. A high rate of mental illness has also been recorded there.
- *Schwarzenburg, Switzerland:* Residents living near a shortwave transmitter have complained about sleeping disorders, lowered concentration, high blood pressure, and anxiety.
- *Skrunda, Latvia:* At a 1994 conference held in Latvia, scientists reported on adverse health effects in humans and animals, as well as tree damage, from a massive Russian space probe and detector-array radar complex in the town of Skrunda. A significant amount of damage to tree growth was correlated with RF power density, as was chromosomal damage in cows who grazed near the complex. Nearby residents complained of headaches and sleep disturbances. Increases

in white blood cells were found in adults and a decreased lung function was identified in exposed children.

- *Zagreb, Croatia:* Long-term exposure to RFs/MWs by air-traffic radar operators and radio-relay-station operators was found to cause significant changes in blood chemistry, electrical brain-wave activity, changes to the blood vessels in the eye, and increases in several forms of cancer.
- *South Patrick Shores, Florida:* A cluster of Hodgkin's disease and ALS has been identified in residents living between three-quarters of a mile and two miles of an air-traffic radar facility.
- *Vernon, New Jersey:* The Centers for Disease Control has identified an unusually high cluster of babies born with Down's syndrome and other genetic abnormalities in Vernon, which has a proliferation of microwave satellite uplinks. Follow-up of this investigation was cut short, and accusations of "dirty politics" have followed.
- *Witchita, Kansas:* Residents in neighborhoods exposed to air-traffic radar from two airports were found to have a higher cancer incidence than in nonexposed neighborhoods.

Other cancer clusters near transmission sites have been reported around Portland, Oregon, and in two communities in California. We need far more research on this, especially epidemiological studies near RF/MW towers. As was true of many of the power-line studies, RF/MW frequencies have been found to cause changes in important biological parameters, even in extremely small amounts. In fact, there may be no safe exposure levels for some frequencies. Research using vanishingly small power levels on yeast cells commonly used in laboratory studies found frequency-specific windows, as if the cells were finely tuned to certain microwave bands in particular. In addition, recent research from both America and New Delhi, India, has found, in studies with mice, DNA damage from low-power microwaves—at the level currently considered safe by the World Health Organization and the FCC.

How to Recognize Antennas in Our Midst

Antennas come in a dizzying variety of sizes and shapes, not to mention the kinds of towers, short and tall, that they are mounted on. So how can you tell what you are looking at? Even for the professional it's not always easy, but here are some general guidelines.

AM, FM, and TV broadcast antennas are usually very tall by law, since they require much more power to function than do local police transmitters, for instance. Those located in remote areas at higher altitudes are probably FM radio and TV transmitters. AM transmitters are typically located in valleys. Police dispatch towers come in varying heights, too. State police systems, which must cover a wide area, require more power and so are taller. Local police dispatch towers are often located close to the police facility and do not need as much power. Public utilities often have their own internal data-communications networks, and these can be substantial towers using microwave frequencies. Most communities also have paging services, which are lower powered but nevertheless must reach about the same area as a local police transmitter, and sometimes the two facilities are combined at the same site. Often hospitals and other municipal buildings lease out space for such transmitters and receivers. If you see an antenna in a neighbor's yard, it is probably for a ham radio system.

Cable TV facilities, which often use satellite transmission, typically are areas filled with parabolic dishes, some ground mounted, some pole mounted. Some are receivers; some are relay dishes that retransmit signals to other areas.

Satellite uplinks are another matter. They transmit signals from the earth to satellites in space. Often they are large ground-mounted dishes that require a lot of power and frequently have substantial leakage and radiation lobes around them. Such uplinks are used by TV stations, telephone companies, private industry, and the entire Internet system, among others. (The

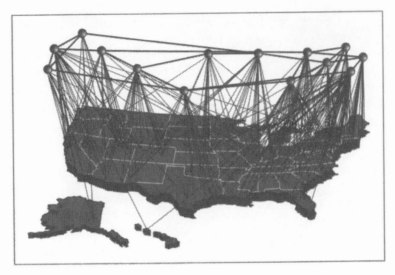

Electromagnetic Traffic on the Internet (approximate uplinks)

illustration shows the Internet system.) Do not confuse the satellite dish on your neighbor's property for a transmitter. Most of these backyard dishes are only receivers, which merely collect transmissions from broadcasting satellites. They are very low powered—but it is still a good idea to stay away from the front of the dish.

Military installations are off-limits to the public, but you might be able to see from a distance tall towers supporting any number of different-shaped antennas and dishes. They are a combination of equipment for everything from dispatch signals, to military communications over long distances (to submarines, for example), to radar, to a national-security communications system used in the event of a massive disaster. (The military is also experimenting with mammoth RF transmitters to "kick the ionosphere" in order to see what happens.) Military communications operate at any number of restricted frequencies, from ELF all the way into the upper microwave bands. The military is also using even higher frequencies, above microwave and infrared, for purposes about which very little information is available.

The size and shape of an antenna and the height at which it is mounted depend on what is being broadcast and the area it is intended to reach. An individual antenna can sometimes physically reflect the wave it is producing—the shorter the wave, the smaller the antenna might be and the higher the frequency on the electromagnetic spectrum. A microwave antenna will be very short in length, compared with an AM antenna. (Microwave transmitters also come in dish form.) Small antennas mounted on taller towers are usually higher frequency, requiring more power (such as for TV broadcasting). But there are exceptions. Sometimes frequency waves are created in "pieces" (for lack of a better word), so what are called half-wave or quarter-wave antennas may be used to create some of the lower FM frequencies, too. Transmitters in dish form often use their rounded contours to help increase the power of a signal before directing it outward through the horn in the center.

In addition, there are such antenna transmission principles as gain (power amplification to increase the transmitting or receiving power of the antenna); impedance (an energy value that must be matched between the antenna and other equipment for good transmission to occur); polarization (the plane of the antenna elements in relation to the earth—either horizontal or vertical); radiation angle (the angle at which the signal radiates to and from an antenna in relation to the horizon); front-to-back ratios (the measurement of power radiated or received from the front of the antenna in relation to power from the back); and radiation pattern (the way waves propagate from an antenna, based on forward gain and front-to-back ratios).

There are literally thousands of antenna shapes and sizes, many of which are used to transmit the same frequencies, and many transmitters are receivers as well. If you live near an antenna of some sort and want to find out what it is, inquire at your local zoning department to see what is on file for that site. Also, many radio stores that cater to ham operators have introductory books that show different antenna structures, but keep in mind that each manufacturer has its own style. Bring a photo if possible, and ask a salesperson to help you identify it.

In general, the antennas in your midst will include radar (air traffic and weather); radio and TV towers (AM, FM, shortwave, police and fire, paging services); satellite dishes (in various sizes, both ground mounted and tower mounted); special relay communications towers for internal utility data networks; and—by far the fastest growing in numbers—cellular-phone towers.

Cellular-Phone Towers

Cellular-phone towers warrant a special section here because just about every community in the country has seen at least one of them appear within the last few years, and there are many questions about their safety.

A few years ago, the FCC, following the direction of Congress, decided we should have a seamless cellular-phone capability from coast to coast. The goal was for a person to be able to get in his or her car at the northernmost tip of Maine, drive to the southernmost tip of California, and have wireless phone conversations the whole way. That no one was demanding this capability other than the companies that stood to profit didn't seem to matter. The philosophy was "if we build it, they will come."

It was obvious that with a massive new communications network looming on the horizon, the big communications companies (AT&T, MCI, U.S. Sprint) would be the main players. So the FCC, again following the lead of Congress, decided to help stimulate the economy by deeming two cellular carriers per district. (It had already divided the country up into cellular-phone regions.) That meant twice as many towers to serve a particular region, and twice as much RF/MW radiation spilling over it.

The agency then put the second licenses into a lottery. All a company had to do was file for one, and a certain number of winners were guaranteed. Word spread quickly through the venture-capital networks, and speculators came from everywhere, most with no communications experience whatsoever. Real-estate speculators on one side of the country, for instance, won remote territories on the other side and immediately resold them at a profit in the millions of dollars, without having made an investment of any kind. The FCC quickly figured out what was occurring and set some guidelines, such as requiring a licensor to erect at least one tower within a year or lose the territory. (Within a short time, most of the territories were in the hands of a few secondary carriers, like Cellular One and MaCaw Cellular, and the big companies were able to merge with them anyhow.)

Towers started going up all over, usually in communities that had no zoning regulations on the books for radio towers, and no expertise in the health concerns that had been raging over microwave radiation for the past four decades.

Communities have been caught by surprise, with too little time to gather information, and so literally thousands of cellular-phone towers now dot the country, many in close proximity to schools and residential neighborhoods, with more being approved every day. This is not to mention the antennas mounted directly on buildings in every city. Or the fact that the former head of the FCC (under the Reagan and Bush administrations, which ushered in all this activity) is now a consultant for the telecommunications companies.

A telecommunications juggernaut appears to be riding roughshod over everyone, including many government agencies. The FCC recently sold entire frequencies in the gigahertz range to the major telecommunications companies, which are developing personal communications systems (PCS)—wireless networks for everything from computers, faxes, and modems to interactive TV and phones. This is being called the "unwiring" of America. Today's cellular-phone tower will undoubtedly become tomorrow's PCS transmitter. That partly explains the rush to site as many cellular-phone towers as possible. The telecommunications giants are after the cable TV companies' markets.

But there are substantial problems, not the least of which are health concerns. Cellular phones transmit in America between 800 and 900 megahertz, which is in the microwave band. The towers have to be matched to the same frequency and are tied to a grid of other towers. Towers require far more power output than a hand-held phone does. As of this writing, at least a dozen lawsuits have been filed regarding brain tumors in cellular-phone users, and suits from those living near cellular-phone towers or transmitters are likely to follow. And the personal communications systems will transmit in the upper-microwave gigahertz ranges. While cellular transmitters are often mounted on huge towers, PCS antennas can be considerably smaller, and hundreds can be scattered throughout a city. Some will be small enough to mount on window ledges or lampposts.

All of this is happening at a time when we have no adequate federal standards for RF/MW exposures; regulatory agencies are in disagreement over a number of key issues; health research is turning up more and more adverse information; fiber-optic cable, which creates very little in the way of EMFs, is being shunted aside as the dominant information-carrying technology; and presumptions of safety are being erroneously made by both the providers and the users of wireless devices.

How Cellular Towers Work

The term cellular, which scientifically refers to a biological process, may come back to haunt this industry. The term is being used to describe wireless-phone service areas, which have been divided into small areas, or "cells," that reuse frequencies in other, nonadjacent "cells." Each "cell" has its own base station, consisting of antennas, radios, and switching equipment.

Cellular towers are the base stations for cellular phones and mobile car phones. They are called cell sites and function as microwave relays, handing off a user's phone conversation to the next tower when the user travels beyond the initial tower's transmission ability. Antennas typically pick up signals from wireless phones and are linked with both wireless and wired networks that route calls from one place to another. The networks are connected directly to local telephone companies; this is what enables wireless phone calls to be placed to conventional phones.

As said before, the cellular-phone network functions at between 800 and 900 megahertz in the microwave bands, and it is frequency modulated at a power output of around 100 watts per channel. Cell sites can typically have many antennas, and their combined power output increases accordingly. Some systems have the capacity for over 100 channels at a single site. In addition, many cell sites are undergoing rapid expansion as users increase. Channels can be split to increase capacity, and power output then increases accordingly.

This is important to understand because often, when cellular-phone company representatives present their information to the public, they speak of the "low power" of cellular towers and try to liken them to the 100-watt light bulbs in every home. Don't be fooled. This is ultra high frequency microwave radiation, not the 60-hertz frequency of your household power supply. Each channel may be 100 watts, but there are many channels, and they can grow exponentially—without any permission from the community, unless it has tight zoning laws.

A typical start-up cell uses omnidirectional antennas that radiate horizontally in a radius of 12 to 16 kilometers, or sometimes more. Cells in remote areas typically need more power to cover a larger distance. The antennas are often mounted to a triangular platform on top of a freestanding pole (monopole), but they can also be mounted on a water tank, a chimney, the side of a building, a rooftop, or anything else tall enough to guarantee proper coverage. Poles can be a lattice type or a tapered steel column, sometimes supported by guy wires. The antennas themselves are several in number and are about 4 meters in length. As many as sixteen transmitters can be connected to a single antenna; the transmitters are usually mounted on top of the platform, with the receivers slightly below. When cells are divided, other directional antennas, approximately 1.2 meters high, are attached to the face of the triangular platform or similar structure.

The radiation patterns from each channel form loops of varying lengths, as the illustration shows. If you live on an adjacent hillside, your exposures can be high, as can exposures near the base of the tower. Of particular concern are roof-mounted antennas, which are often just a few feet above head height. Anyone near such a transmitter may experience extremely high exposures, as will those in adjacent buildings.

A 1992 study conducted by Petersen and Testagrossa, at the radiation protection department of AT&T Bell Laboratories, found that population exposures in the vicinity of several pole-mounted cellular towers were less than 100 μW/cm^2 per channel. For roof-mounted systems, near-field exposures were found to be

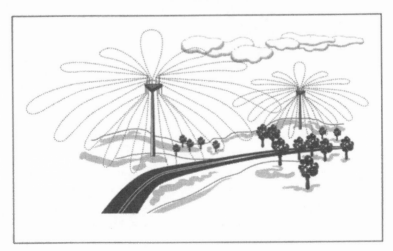

Radiation Pattern of Cellular-Phone Towers

less than 30 W/m² at 0.7 meters from the antenna. They concluded that such exposures were within safe limits according to the thermal-based ANSI/IEEE C95.1 standard used by the FCC. However, as stated earlier, this standard is meaningless in the face of the nonthermal bioeffects discovered in the last several years.

For a thorough discussion of the ANSI standard, its history and where we are today, see Chapter 2. It's an important chapter for communities that want to protect themselves against cellular or other transmission facilities and are puzzled by what's happened at the federal level.

Standards

Public pressure is mounting over RF/MW exposures, the same way as it has over the ELF power-line frequencies. Communities have prevented cellular towers from being sited; radio and TV broadcast towers have had leases revoked or not renewed; and some communities have moved to adopt more stringent regulations than those required by the FCC.

For over ten years, there has been virtually no federal funding for civilian RF/MW medical research. That situation needs to change quickly before the new round of wireless-communications technology washes over us. A $65 million EMF research and information dissemination program (called the RAPID program) was ordered by Congress in 1992, but its efforts are mainly concentrated in 60-hertz and ELF research. A similar program for the RF/MW ranges is not only imperative, but decades overdue. Industry appears to be trying to do its part. For instance, the Cellular Telecommunications Industry Association (CTIA), after adverse publicity concerning the possible causal link between brain tumors and cellular phone use, voluntarily set aside $25 million in a blind trust for research into the bioeffects of those frequences. Sensitive to consumer suspicions about the outcome of any industry-sponsored research, the CTIA established the blind trust so as to avoid accusations of grantor bias. They also enlisted the Harvard University Center for Risk Analysis to conduct peer review of any of the studies that CTIA funds. To their credit, CTIA, which is the umbrella organization for the cellular phone industry, did not wait, as the RAPID program did, for Congress to mandate a research program and require that industry match the funds. To its discredit, the CTIA's Science Advisory Group (SAG) is comprised mainly of researchers who are known to believe that exposure to RF/MW energy below subthermal thresholds does not cause any adverse biological reactions — views similar to those of the original groups who first looked at the subject in the 1950s and 1960s. One hopes that we are not destined to repeat the mistakes of the early researchers described in Chapter 2.

Industry-sponsored research is not automatically tainted like many consumers think. Industry *should* sponsor its own research since it creates the products and devices that need to be regulated. But such research must be peer-reviewed by others with no stake in the outcome other than scientific knowledge, and it must not be the only research that is done. Government must conduct independent research as well, with an eye toward protecting consumer health, not industry dollars. Toward that end, the EPA has recently revived its nonionizing electromagnetic radiation (NIER) program, and has commissioned a two-year study by the National Council on Radiation Protection and Measurements (NCRP) concerning human exposures to weak, modulatd RF/ MW radiation. The EPA has also gone back to its efforts to set exposure guidelines, efforts that were halted under the Reagan administration. It is essential that the EPA be allowed to carry out these efforts and that it not be politicized by another business-friendly administration.

As said in Chapter 2, the National Council on Radiation Protection (NCRP) — the only agency mandated by Congress to set radiation limits for a range of products and devices — has made the most stringent recommendations for general exposure to the

RF/MW frequencies. Here is a comparison of the ANSI and the NCRP standards: The 1982 ANSI guidelines (currently adopted by the FCC) for RF radiation between 300 kilohertz and 100 gigahertz are frequency dependent (meaning they recognize that the human anatomy absorbs energy at some frequencies more efficiently than at others) and restrict the frequency range of from 30 to 300 megahertz to maximum exposure levels of 1 mW/cm^2 as averaged over a six-minute period. Levels can exceed that for shorter periods. These recommendations were based on a determination that the threshold for hazardous biological effects (meaning tissue heating) was around 4 watts per kilogram (4 W/kg) of energy absorbed by the body (the specific absorption rate, or SAR). A safety factor of 10 was incorporated, and the guideline for the SAR threshold became 0.4 W/kg. This standard makes no distinction between occupational exposures and long-term continuous exposures.

An updating of this standard now under consideration sets a two-tiered recommendation for the general population and for occupational exposure (called uncontrolled and controlled environments), but it is still based on the thermal-effects-only presumption. This current revision is also more lenient in some circumstances than the 1982 version that the FCC now uses—a clear step in the wrong direction—and it should not be adopted. (The standard has become so complex that many broadcasters have difficulty figuring it out—let alone the lay public.)

For the frequency range between 300 kilohertz and 1,000 megahertz (1 gigahertz), the ANSI standard excludes radiating devices with an input power of 7 watts or less, which includes cellular phones. These exposure limits may be exceeded if exposures produce specific absorption rates below 0.4 W/kg as averaged over the whole body, or below 8 W/kg if averaged over any one gram of tissue.

The NCRP recommendations differ only slightly from the ANSI-1982 standard. They are virtually the same for occupational exposures because it is presumed that professionals understand and consent to the uncertainties of such exposures. The NCRP, however, recognizes that the general public rarely understands or consents to such exposures, and that the sick, the old, the very young, and pregnant women and their babies are not factored into professional exposures. The NCRP therefore recommends a standard for the general population that is five times more stringent than ANSI-1982 at a level not to exceed 0.08 W/kg, but this is over a thirty-minute exposure. (While the absorption is presumed to be less, the length of time allowed is longer.) The NCRP also acknowledges the existence of nonthermal effects and takes some modulated frequencies into consideration for both the general population and for workers.

Other recommendations of various organizations are based on versions of these two standards. That of the International Radiation Protection Association (IRPA) is similar to the NCRP recommendations in that a greater degree of protection for the general population is suggested than for workers. And the American Council of Government Industrial Hygienists recommends basically the same as the ANSI standard.

The EPA has requested that the FCC adopt the NCRP standards over the newly revised ANSI guidelines now under consideration by the FCC. (By the time this book goes to press, a decision may have been made. Watch your newspaper or phone the FCC.) The EPA has also asked the NCRP to update its 1986 standards, and it has agreed to do so.

There are currently no federal standards for exposure of the general public to RF/MW radiation generated from over 4,500 AM stations, 4,400 FM stations, 1,100 TV stations, and thousands of cellular-phone and other transmitters. It is generally agreed that the EPA has the responsibility for setting such standards, and it attempted to do so in 1986 but got sidetracked. The agency has now gone back to work on them. Many of the 1986 recommendations are expected to be repeated. They included an intent to set limits "as low as reasonably achievable"; transmission power reductions; the regular monitoring of transmission sites; taller antennas and the replacement of old antennas with more efficient equipment; requirements for fencing, larger setbacks, and a number of other things. Many of these recommendations are already being considered by the FCC.

In the absence of federal standards, states and municipalities have enacted regulations of their own. Some states, like Connecticut, adopted the ANSI standard intact, thinking that it would give at least some protection. Massachusetts, on the other hand, adopted the IRPA recommendations with a two-tiered occupational and general-population exposure (200 μW/cm^2 for whole-body average SARs to 0.04 W/kg for frequencies above 3 megahertz). Puerto Rico adopted the NCRP recommendations and tried to enact even more stringent ones, but met with sharp opposition from the cellular-phone industry. Portland, Oregon, adopted standards more stringent than the IRPA recommendations, at 100 μW/cm^2, as well as several zoning requirements pertaining to transmission towers, including the regular measuring, reporting, and monitoring of all EMF emissions. Oldham County, Kentucky, has set standards for communications facilities, on nonionizing radiation exposures to the general public, at .2 mW/cm^2. And Seattle, Washington, has height and emissions regulations.

Even some U.S. Air Force bases have set their own stringent standards, as have some government contractors and research labs. The Kirkland Air Force Base in New Mexico, noting the abundance of research on nonthermal effects, instituted exposure limits to 100 μW/cm^2 for the 30 megahertz to 100 gigahertz range. This is similar to the limits in effect at the Johns Hopkins University Applied Physics Lab for researchers there. By contrast, the ANSI-1992 standard allows occupational exposures 10 times greater, between 100 and 300 megahertz, and up to 100 times greater above 3 gigahertz. Hughes Aircraft has also moved to protect employees with standards more stringent than ANSI's.

The ability of local communities to set more protective standards may soon stop. The cellular phone industry recently asked the FCC to override all local jurisdiction so they can site an estimated 100,000 new cellular towers by the year 2000 without local interference. The recent telecommunications deregulation bills in both houses of Congress had siting preemption clauses inserted to override local control. Industry wants to make the ANSI standard the law of the land. Contact your government representatives to help ensure local control.

What a Community Can Do

It is better for local communities to have something on the books rather than nothing. Broadcasters and cellular-phone companies usually object, but they, too, would be better off dealing with clear-cut regulations than confronting angry, fearful citizens at every local meeting. In the absence of *good* federal standards, what other choice does a responsible municipality have?

The politics of radio towers—especially cellular-phone towers—has become fascinating in the last several years. Because there is so much citizen opposition to new towers, many communications companies have been combing through old records to locate obsolete tower sites and reactivate them. (If you live near the site of a former radio tower, make sure the tower is really inactive according to time-limit zoning regulations.) Another tactic of the companies is to promise a town free police or fire department radio service, or to promise schools free computer hookups or radio stations. Sometimes their route is through the state police and emergency broadcasting facilities; the companies help to site new facilities for those services and then piggyback on them with cellular transmitters, paying a rental fee to the state. Sometimes a cellular-phone company has resorted to searching town records to find someone who is in arrears on property taxes and approaching that person to lease the land at a substantial fee. In such cases it is important for landowners to know that they can be named, along with the leasing company, in lawsuits for health damages or the devaluation of nearby property.

As a federal agency, the FCC can preempt both state and local regulations, but it has thus far been reluctant to override more stringent local regulations. (This may soon stop.) Some states have siting councils that can preempt local zoning in the placement of public-utility projects and radio towers. In fact, many communities have argued about whether cellular-phone towers can be considered public utilities. Some states say yes (New York); others say no (Connecticut). In states with siting councils, citizens have an additional layer of government to address—one that too often has a cozy relationship with the very industries it is supposed to regulate. But even in states with siting councils, local communities can enact more stringent legislation. They just have to be prepared to battle it out in court, if need be. (Those who adopt the NCRP standards are on firm legal ground, since it is the only organization with a congressional mandate to set regulations.) In states with local jurisdiction, citizens have the best chance to enact productive guidelines, sometimes through zoning, sometimes through the health department, sometimes both.

Here are some things for you and your local citizens' groups to consider:

- Get emissions standards adopted, either the NCRP regulations for RF/MW exposures or the more stringent standards used at Kirkland Air Force Base. Contact the NCRP at 800-229-2652 and ask for Report No. 86, *Biological Effects and Exposure Criteria for Radiofrequency Electromagnetic Fields.* The book costs $50 and can be ordered by phone with a credit card. (This report is in the process of being updated; find out when the new version is due.) Keep in mind that although this is one of the most protective standards available for population exposures, it is based on energy-absorption models that may prove to be erroneous or that will be ineffective under certain circumstances. The amount of microwave radiation produced from cellular towers alone should not be underestimated, let alone that from other transmitters. Cellular phones, which are much lower powered than the towers, can emit between 3 and 6 watts of power, which in about fifteen minutes can create enough heat to cook a hot dog. A tower emits 75 to 100 watts per channel and can have over a hundred channels.

As a general guide, some of the thresholds for bioeffects noted at different power densities are: *

Exposure (mW/cm^2)	Effects
100	Threshold of cataract formation
28	Teratogenic and tumor-causing effects
10	Molecular and genetic effects (thermal)
4	Threshold for neuroendocrine effects
0.05	Decreased sperm counts
0.03	Increased brain-amine levels
0.01	Altered brain permeability

*From W. R. Adey, "Tissue Interactions with Nonionizing Electromagnetic Fields," 1982.

Also, low-level microwave radiation has been found to alter certain drug absorption rates in humans, especially for glaucoma medications. Proximity to hospitals and nursing homes should be considered.
• Contact the zoning office in Portland, Oregon, and request a copy of its radio-frequency regulations. Oregon has put more thought into this subject than perhaps any other state.
• Some states and municipalities have adopted tower-sharing laws stating that no new facility can be sited if a facility already exists to service the area. It requires one carrier to lease space from another provider and is intended to stop the needless proliferation of towers.

- Decide whether cellular towers are a public utility or not. In many municipalities, a public utility is allowed by right in every zone and can be established without a public hearing, whereas radio and TV towers cannot. Several communities have enacted zoning amendments that include cellular sites in existing radio and television tower classifications. If your town does not have such regulations, get them passed quickly. Pensacola, Florida, amended its zoning laws to define a commercial communications tower as "a structure situated on a non-residential site that is intended for transmitting or receiving radio, television, or telephone communications, excluding those used exclusively for dispatch communications."

- Some communities have minimum setback laws that are specifically related to tower height. Some specify six times the height of the tower, others four times the height. Other communities require that towers comply with regular zoning setbacks, which are often inadequate for towers. The general rule should be for larger, not smaller, setbacks. Also, many communities require fencing to restrict public access to tower sites but exclude barbed wire and razor wire in residential areas.

- Many communities specify that obsolete or unused facilities must be removed within twelve months of cessation of operations at the site.

- Multnomah County, Oregon, established distances for cellular-phone antennas: at least six feet from the surface of any habitable structure, and a height of at least fifteen feet from the nearest exterior surface.

- Many communities have grouped communications facilities in one area, rather than allowing them to be scattered throughout the community. In general, this is a good idea if it also requires enough space around the area—it should be considerable—to allow for future growth. Such areas are sometimes referred to as antenna farms. The electromagnetic environment, even several miles away, can be extremely complex.

• Encourage satellite-communications technology rather than ground-station transmitters. Satellite communications remove the exposures to a higher layer in the atmosphere (and therefore away from people), and when the signals return to earth they are typically disseminated through wired systems, like fiber optic cables, rather than through additional wireless channels. But be sure that the areas around satellite uplinks are large enough, are well monitored, and are scrupulously measured for emissions. There can be significant side lobes of radiation from uplinks.

• Require the regular measurement, monitoring, and reporting of exposures from all transmission facilities, including those of police and fire departments and paging services, by the transmission facilities themselves, but have the reports verified by an independent testing company. Because wave propagation can be complex, depending on such factors as frequency, power output, nearby structures, topography, conductive materials like metal water towers, and unexpected couplings with other frequencies from power lines or other radio towers, a handful of communities have enacted a specific test protocol that was developed by a bioengineer and health physicist, Carroll Adam Cobbs, of Seattle, Washington (see accompanying table).

The Cobbs protocol is designed to test various aspects of the ambient background before a tower goes on-line and again afterward. It is intended to provide a crucial piece of information for future epidemiological studies that has been completely lacking in the past—the precise environmental alteration at a set moment in time from which health effects might be observed. Existing towers can be shut off and the environment tested; then the facilities can be turned back on and the environment retested. This will give a community precise knowledge about what the RF/

THE COBBS TEST PROTOCOL*

Test Modality	Suggested Equipment
Microwave power density measured at several locations	HP 437B Power Meter with 8542A Thermistor sensor and calibrated horn antenna, to include SWR measurements at each setup
RF field density using broad-band survey meter	Holaday Instruments RF Survey Meter with Isotropic Broad-band Probe
Broad-band spectrum analysis covering the range from 50 Hz to 2.9 GHz	HP 35665A (Low-Freq.) and HP 8591A or HP 8560 (VHF to microwave)
Carrier noise measurement	HP 11729C Carrier Noise test set
Transmitter test and RF signal characterization Also Analyzer to record modulation characteristics of the detected signals	HP 8901A Modulation
Frequency and time-interval analysis to graphically illustrate frequency and time and phase interval information	HP 5372A Freq. and Time Interval Analyzer

*Printed with the permission of Carroll Adam Cobbs, M.S. For additional information, contact him at 1011 Boren Ave., #184, Seattle, WA 98104; phone: 206-248-2336.

THE COBBS TEST PROTOCOL

(continued)

Test Modality	Suggested Equipment
60 Hz electric and magnetic field strength	Holaday Instruments low-frequency measurement system with fiber optic isolation cable and accessories *or* Combinova low-frequency magnetic field meter with fiber optic probe isolation cable (60 Hz)

MW radiation is, but measurements can be tricky and specific equipment is required.

Suggested Methods and Data Collection

- All tests must be conducted both before tower construction or operation and after the system is functional.
- All possible antenna configurations and power outputs must be examined.
- Tests should be conducted during normal times of day with other EM sources at normal operation.
- Note all "hot spots" (areas or "nodes" of high local intensity) and "nulls" (areas where fields may be sharply diminished or cancelled). A wide-area "map" should be prepared that encompasses the normal activities of the population which will be subjected to the exposures. This map should then become the template for all subsequent studies.
- The exact coordinates of each test location should be recorded on copies of the map. The establishment of this map can be facilitated by using a commercially available, hand-held Global Positioning Satellite System (GPSS) receiver. This unit will provide the most accurate land coordinates ir-

respective of structures and boundaries which may change with time.

- The orientation of each antenna or probe should (as much as possible) be the same, and readings of minimums and maximums should be noted. Wherever possible, isotropic antennas and probes should be used.
- All parameters under test should be referenced to the same coordinates at each test location.
- Measurements should be taken at locations and at heights above ground (or other surfaces) where individuals may reasonably be found.
- Measurements should not be made within 20 centimeters of any object.
- All raw data should be stored by test site, date, and modality on digital media as well as by written notebook entry. This will permit much of the equipment specified (spectrum analyzer, etc.) to remain at an analysis center where the probe data may be reviewed. It will also permit retrospective analysis of the data by others. Most of the equipment specified in the protocol has this storage capability.
- All equipment should be calibrated regularly by an independent facility with traceable calibration sources.
- All test team members should show proof of training on each piece of equipment as well as have familiarity with electric and magnetic field measurement procedures.
- If epidemiological studies are considered at some point in the future, this data may be readily used in correlation studies with demographics, medical histories, etc.

The precise measurement of RF/MW radiation is just now becoming technically feasible. Better equipment has been developed in the past few years, in addition to a better understanding of the subtleties of its use. Measuring nonionizing radiation in the field has become a fine science and is best done by professionals. Equipment is typically expensive as well as finely calibrated;

it requires special expertise and training to use and interpret. (A simple gaussmeter won't suffice.) Because of the absence of federal funding, it would not be unreasonable to ask the transmission companies to pay for equipment and testing. Mr. Cobbs has specified certain equipment because of its known technical quality and to achieve particular test parameters.

It has become obvious to many communities that, with such a large regulatory gap at the federal level, local protection is crucial. But few communities have had the expertise to write regulations that are truly effective. A community wanting to do something quickly might consider writing into its zoning or health regulations the NCRP guidelines, along with the Cobbs Test Protocol, which can also be made retroactive to existing towers. Broadcasters in your area will then know exactly what is required of them. Many communications companies that initially have grumbled about such restrictions have come to respect the communities that hold public safety in high regard, and they have also come to welcome the opportunity to learn more about the effects (or lack thereof) that their technology has on an environment.

If you are a private citizen living near a communications system of some kind and would like to take measurements, the best thing to do is phone a professional EMF-testing consultant. Unfortunately, the equipment to accurately measure the radio/microwave frequencies is expensive, and no consumer models (like the simple gaussmeters for measuring 60-hertz fields) are yet available. As consumer curiosity about RF/MW rises, such equipment will undoubtedly become available.

Some people have experimented with inexpensive wave guides, but their calibrations can be confusing for the nonprofessional. A simple radar detector will clue you in to that frequency in your immediate environment, but professional testing is best. If you are experiencing RF interference with your radio or TV reception, the FCC will sometimes make a field visit to take measurements. It's always worth a call.

CONCLUSION

WE HAVE COVERED an immense amount of information in this book, spanning centuries of inquiry by some of the great minds of each millennium, in all branches of science. Whole volumes have been written on any number of the areas we've merely touched upon here.

The nonionizing band of the electromagnetic spectrum will probably turn out to be far more significant than anyone heretofore imagined. There is the distinct possibility, for instance, that entrainment phenomenon, resonance relationships, and other reactions to nonionizing electromagnetic fields will prove to be a critical, but hidden, variable in *all* scientific research, be it experiments with animals, environmental observations in the field, laboratory experiments using culture specimens, or epidemiology studies tracking disease states in the population.

Such a variable would touch all research, across all disciplines, not just EMF-related research. In a laboratory setting, for instance, what if animal research that found positive carcinogenic associations with benzene was actually influenced by the presence of overhead fluorescent lights, or an electrical switching cabinet adjacent to the laboratory? What if some of the finest research in molecular biology, providing answers today about genetic defects that are unparalleled in medical history, is in fact

measuring subtle DNA alterations in cell samples caused by proximity to RF/MW transmission facilities? What if all of our otherwise painstakingly careful research protocols are slightly tainted
by such a ubiquitous variable that it goes virtually unnoticed? It
would mean that everything we think we understand scientifically
is askew.

In the environment outside the lab, what if the spraying of
pesticides turns out to be more detrimental to a farming community at a certain time of day because of changes in human circadian rhythms that allow trace amounts of chemicals to be more
efficiently absorbed at that time? And what if we could more effectively use those pesticides to kill insects that might be more or
less sensitive to chemicals depending on their own biorhythms?
Or what if the positioning of high-tension corridors in relation to
the earth's ley lines counts in either a positive or negative way to
those living nearby? Or what if microscopic toxins, bacteria, and
viruses can hitchhike a ride on some of the RF/MW frequencies
(which pass right through body tissue) and gain easy access to
the interior of the cell through the cell's protective membrane,
effectively giving them an edge over other species' evolutionary
defenses? These are only a hint at the kinds of questions that can
be raised regarding this issue; there are thousands more. We are
just beginning to understand the implications of some of the research cited in this book.

The global population, human, animal, and vegetation alike,
is an unwitting test subject as many of these EMF technologies
continue to proliferate, especially with the increased use of the
RF/MW bands. But are all electromagnetic fields detrimental to
us? Most likely not. Are some more detrimental to some species
than to others? Undoubtedly so. Can we all, perhaps, tolerate a
certain amount of exposure from all frequencies across the spectrum, but not a concentration of specific frequencies in certain
species? Probably. But this remains to be seen, and then the question arises, how do we decide which species are more or less valu-

able to the ecosystem? Does our craving for the newest technology supersede the right of other species to exist?

A host of competing societal rights, and even some constitutional issues, will have to be addressed as well. For instance, does the federal government have the right to override states' rights in siting communications facilities when there are no federal safety standards in place to protect public health? Does the FCC have the right to auction off our finite national airwaves to the telecommunications giants? Does the military have the right to withhold important research from the general public (who funded it in the first place) or continue to conduct itself in near-complete secrecy regarding what it knows, and is doing, with electromagnetic field research? Can communities set their own safety standards without state or federal interference or ban some installations altogether? Do businesses have the continuing right to bring products, devices, and medical tools into the marketplace without even a minimal understanding of some of the more pertinent EMF bioeffects research? Should consumer products be labeled for EMF emissions? Should we begin to systematically shut down communications facilities and test ambient backgrounds without them in operation, in order to better understand precisely how each installation is altering its surrounding environment? And should we be far more judicious about allowing new ones to go on-line?

In the larger EMF discussion, we will need to bring a sense of humanistic values back to the table and rely less on scientific experts as our only guides, especially where this intersects with the role of setting exposure standards. Scientists are concerned with observations and probabilities and are not trained to understand the societal implications of every facet of their work. Scientific research is about understanding natural processes and should not be concerned with setting standards per se. That's our job as citizens and lawmakers. We do not need a finite understanding of underlying physiological mechanisms to set standards. What we need is the *appropriate* scientific information upon which to make those decisions. And we desperately need to fund new research, rather than just reanalyze the same old thermal-based studies.

A whole new EMF era is dawning with virtually no safeguards in place and with all these myriad questions unanswered. "Wireless" America is looming on the horizon. It will alter our ecosystem in a way never experienced before. The stakes may be higher than we know at this juncture. Or they may turn out to be lower than a summation of the research contained in this book indicates. But erring on the side of caution has never proven an ill-advised course of action and is certainly a more intelligent approach than the reckless abandon with which we have thus far embraced many modern technologies.

It will take a combined nonpartisan effort on the part of government, business, private citizens, and the scientific community if we are to right the wrong course that it appears we are on. This may seem like an impossibly naive notion, given today's political climate, and one that has had little success to date. But hope springs eternal . . .

So much more is known today about EMFs than just a decade ago; the public is far more educated about the subject than it used to be; and the time has certainly come to pull the divergent pieces of the picture together and see it for what it is—a potentially titanic problem.

SELECTED BIBLIOGRAPHY

Chapters 1 and 2

Brodeur, Paul. *Zapping of America*. New York: W. W. Norton & Company, Inc., 1977.

Duchene, A. S., Lakey, J. R. A., and Pepacholi, M. H., eds. *IRPA Guidelines on Protection Against Non-Ionizing Radiation*. Oxford: Pergamon Press, 1991.

Ericsson, Stephanie. "The Ways We Lie." *Utne Reader*, no. 54, Nov./Dec. 1992, pp. 56–72.

Markoff, John. "A Web of Networks, an Abundance of Services." *New York Times,* Feb. 28, 1993, p. 8F.

Martin, Harry C. "ANSI RF Guidelines Are Reviewed." *Broadcast Engineering,* Jan. 1993.

Microwave News, May/June 1992; Jan./Feb. 1993; March/April 1993; Sept./Oct. 1993; Jan./Feb. 1994; March/April 1994; May/June 1994.

1991–1992 Threshold Limit Values for Chemical and Physical Agents and Biological Exposure Indices. American Conference of Industrial Hygienists, Cincinnati, Ohio, 1991.

Steneck, Nicholas H.; Cook, Harold J.; Vander, Arthur J.; and Kane, Gordon L. "Standards for Microwave Radiation." *Science,* vol. 208, June 13, 1980.

Steneck, Nicholas H. *The Microwave Debate*, Cambridge, MA: MIT Press, 1980.

U.S. Environmental Protection Agency. *Evaluation of the Potential Carcinogenicity of Electromagnetic Fields*. Draft review, EPA/600/6-90/005B, Oct. 1990.

Chapter 3

Adair, Robert K. *The Physics of Baseball*. New York: Harper & Row, 1990.

Becker, Robert O. *Cross Currents: The Perils of Electropollution, the Promise of Electromedicine*. Los Angeles: Jeremy P. Tarcher, Inc., 1990.

Bernstein, Jeremy. *The Tenth Dimension: An Informational History of High Energy Physics*. New York: McGraw-Hill Publishing Company, 1989.

Electrical and Biological Effects of Transmission Lines: A Review. Portland, OR: U.S. Department of Energy, Bonneville Power Administration, 1989.

Miller, Franklin, Jr. *College Physics*, 5th ed. New York: Harcourt Brace Jovanovich, 1982.

Weinberg, Steven. *The Discovery of Subatomic Particles*. New York: W. H. Freeman and Company, 1990.

Zukav, Gary. *The Dancing Wu Li Masters: An Overview of the New Physics*. New York: Bantam Books, 1980.

Chapter 4

American Society of Dowsers, Danville, VT. Pamphlets on dowsing, privately printed.

Becker, Robert O. *Cross Currents: The Perils of Electropollution, the Promise of Electromedicine*. Los Angeles: Jeremy P. Tarcher, 1990, pp. 173–84.

Becker, Robert O., and Selden, Gary. *The Body Electric: Electromagnetism and the Foundation of Life*. New York: William Morrow, 1985.

Blakemore, R. P., Blakemore, N. A., and Frankel, R. B. "Bacterial Biomagnetism and Geomagnetic Field Detection by Organisms." In *Modern Bioelectricity*, ed. Andrew A. Marino. New York: Marcel Decker, Inc., 1988, pp. 19–34.

Electrical and Biological Effects of Transmission Lines: A Review. Portland, OR: U.S. Department of Energy, Bonneville Power Administration, 1989.

Gerber, Richard. *Vibrational Medicine: New Choices for Healing Ourselves*. Santa Fe, NM: Bear & Company, 1988.

Gould, J. L. "The Case for Magnetic Sensitivity in Birds and Bees (Such As It Is)." *American Scientist* 68 (1988): 256–67.

Grove, Bob. "Man: The Human Receiver." *Monitoring Times,* March 1991.

Lanzerotti, L. J. "The Earth's Magnetic Environment." *Sky and Telescope,* Oct. 1988.

Lee, J. M., Jr., and Reiner, G. L. "Transmission Line Electric Fields and the Possible Effects on Livestock and Honeybees." *Transactions of the American Society of Agricultural Engineers* 26(1): 279–86.

Lerner, E. J. "The Big Bang Never Happened." *Discover,* June 1988.

McCleave, J. D., Albert, E. H., and Richardson, N. E. *Perception and Effects on Locomotor Activity in American Eels and Atlantic Salmon of Extremely Low Frequency Electric and Magnetic Fields.* Final Report No. N000014-72-C-0130, Naval Electronics System Command, Department of Zoology, University of Maine, Orono, Maine.

"NOVA: Supersenses." PBS telecast.

Papi, R., Meschini, E. M, and Baldaccini, N. E. "Homing Behavior of Pigeons Released After Having Been Placed in an Alternating Magnetic Field." *Comparative Biochemical Physiology* 76A(4): 637–82.

Chapter 5

Blackman, Carl. "Do Electromagnetic Fields Pose Health Problems?" Colloquium presentation, April 18, 1990. In *Frontier Perspectives,* vol. 2, no. 2, Fall/Winter 1991.

Burr, Harold Saxon. *Blueprint for Immortality: The Electric Patterns of Life.* London: Nevill Spearman, Ltd., 1972.

———. "Diurnal Potentials in the Maple Tree." *Journal of Biology and Medicine* 17: 727–34.

Burr, H. S., and Northrop, F. S. "The Electromagnetic Field Theory." *Quarterly Review of Biology* 10 (1935): 322–33.

Carson, Rachel. *Silent Spring.* Boston: Houghton Mifflin, 1962.

Fenson, D. S. "The Bio-electric Potentials of Plants and Their Functional Significance, I: An Electrokinetic Theory of Transport." *Canadian Journal of Botany* 35 (1957): 573–82.

———. "The Bio-electric Potentials of Plants and Their Functional Significance, II: The Patterns of Bio-electric Potential and Exudation

Rate in Excised Sunflower Roots and Stems." *Canadian Journal of Botany* 36 (1958): 367–83.

———. "The Bio-electric Potentials of Plants and Their Functional Significance, III: The Production of Continuous Potentials Across Membranes in Plant Tissue by the Circulation of the Hydrogen Ion." *Canadian Journal of Botany* 37 (1959): 1003–26.

———. "The Bio-electrical Potentials of Plants and Their Functional Significance, IV: Some Daily and Seasonal Changes in the Electric Potential and Resistance of Living Trees." *Canadian Journal of Botany* 41 (1963): 831–51.

Glinz, Franz. "The Forest Is Dying from 'Electrosmog' (or Electromagnetic Contamination)." *Auto-Illustrierte,* Feb. 1992.

Grove, Bob. "Man: The Human Receiver." *Monitoring Times,* March 1991.

Harvalik, Z. V. "A Biophysical Magnetometer-Gradiometer." *Virginia Journal of Science* 21:2 (1970): 59–60.

Hertel, H. "The Path into the Dying Forest: The Forest Dies As Politicians Look On." *Raum & Zeit Magazine,* no. 51, May/June 1991.

Lohmeyer, Michael. "Environment Contaminated by Microwaves: Directional Radio Beams Cut Breaks into Forests." *Die Presse Independent Daily for Austria,* July 31, 1991.

Lund, E. J. *Biological Rhythms in Psychiatry and Medicine.* Austin: University of Texas Press, 1947.

McAuliffe, Kathleen. "The Mind Fields." *Omni Medical Edition,* May/June 1988.

Polk, Charles. "Biological Effects of Low-Level Low-Frequency Electric and Magnetic Fields." *IEEE Transactions on Education,* vol. 34, no. 3, Aug. 1991.

Tompkins, Peter, and Bird, Christopher. *The Secret Life of Plants.* New York: Harper & Row, 1973.

Volkrodt, Wolfgang. "Are Microwaves Faced with a Fiasco Similar to That Experienced by Nuclear Energy?" *Wetter-Boden-Mensch,* April 1991.

———. "Electromagnetic Pollution of the Environment." *Environment and Health: A Holistic Approach,* chap. 8. Luxembourg: Avebury Press, 1989.

Chapter 6

Benet's Reader's Encylopedia, 3rd ed.

Chopra, Deepak. *Quantum Healing: Exploring the Frontiers of Mind/ Body Medicine.* New York: Bantam Books, 1989.

Clendening, Logan. *Source Book of Medical History.* New York: Dover Publications, 1942.

Dibner, Bern. *Luigi Galvani.* Norwalk, CT: Burndy Library, 1971.

Donden, Yeshi. *Health Through Balance: An Introduction to Tibetan Medicine,* trans. Jeffrey Hopkins, Lobsang Rabgay, and Alan Wallace. Ithaca, NY: Snow Lion Publications, 1986.

Encyclopedia Britannica, 15th ed.

Gerber, Richard. *Vibrational Medicine: New Choices for Healing Ourselves.* Santa Fe, NM: Bear & Company, 1988.

Hartman, Franz. *The Life and Teachings of Paracelsus.* New York: Rudolph Steiner Publications, 1973.

Meyer, Herbert. *A History of Electricity and Magnetism.* Norwalk, CT: Burndy Library, 1972.

Schiegl, Heinz. *Healing Magnetism: The Transference of Vital Force.* York Beach, Maine: Samuel Weiser, Inc., 1987.

Chapter 7

Becker, Robert O., and Selden, Gary. *The Body Electric.* New York: William Morrow, 1985.

Becker, Robert O. *Cross Currents: The Perils of Electropollution, the Promise of Electromedicine.* Los Angeles: Jeremy P. Tarcher, Inc., 1990.

"Biological Effect of Electromagnetic Fields." *IEEE Spectrum,* May 1984, pp. 57–69.

Coghill, Roger. *Electropollution: How to Protect Yourself Against It.* Wellingborough, England: Thorsons Publishing Group, 1990.

Columbia University College of Physicians and Surgeons. *Complete Home Medical Guide,* eds. Donald Tapley, Robert Weiss, and Thomas Morris. New York: Crown Publishers, Inc., 1985.

Conference Notes from Transmission & Distribution Magazine's EMF Conference, 1993. Papers presented by Imre Gyuk, Joseph Kirchvink, and Larry Anderson.

Davis, Lisa. "Clockwise Exercise." *Health Magazine,* Nov. 1991, p. 90.

Dolnick, Edward. "Snap Out of It." *Health,* vol. 6, no. 1, Feb./March 1992.

Gilbert, Susan. "Harnessing the Power of Light." *New York Times Magazine,* April 26, 1992.

Gray, Henry. *Gray's Anatomy,* 1901 edition, ed. T. Pickering Pick. Philadelphia, PA: Running Press, 1974.

Grove, Bob. "Man: The Human Receiver." *Monitoring Times,* March 1991, p. 22.

Konig, H. "Biological Effects of Extremely Low Frequency Electrical Phenomena in the Atmosphere." *Journal of Interdisciplinary Cycle Research* 2 (1971).

"Looking to Biology for the Causes of Crime." *New York Times,* Week in Review, Jan. 23, 1994.

McAuliffe, Kathleen. "The Mind Fields." *Omni Medical Edition,* May/June 1988, p. 15.

"The Mysteries of Melatonin." *Harvard Health Letter,* vol. 18, no. 8, June 1993.

Persinger, M. A., Ludwig, H. W., and Ossenkopp, K. P. "Physiological Effects of Extremely Low Frequency Electromagnetic Fields: A Review." *Perceptual and Motor Skills* 36 (1973).

Polk, Charles. "Biological Effects of Low-Level Low-Frequency Electric and Magnetic Fields." *IEEE Transactions on Education,* vol. 34, no. 3, Aug. 1991.

Sheppard, A. R., and Eisenbud, M. *Biological Effects of Electric and Magnetic Fields of Extremely Low Frequency.* New York: New York University Press, 1977.

Smith, Cyril W., and Best, Simon. *Electromagnetic Man.* New York: St. Martin's Press, 1989.

Sugarman, Ellen. *Warning: The Electricity Around You May Be Hazardous to Your Health.* New York: Simon & Schuster, 1992.

U.S. Environmental Protection Agency. *Evaluation of the Potential Carcinogenicity of Electromagnetic Fields.* Draft Review, EPA/600/6-90/005B, Oct. 1990.

Yale University School of Medicine Heart Book, ed. Barry L. Zaret, Marvin Mosher, and Lawrence S. Cohen. New York: Hearst Books, 1992.

Chapter 8

Adey, W. R. "Frequency and Power Windowing in Tissue Interactions with Weak Electromagnetic Fields." *Proceedings of IEEE* 68:1 (1980): 119–25.

Adey, W. R., Bawin, S., and Lawrence, A. F. "Effects of Weak Amplitude-Modulated Fields on Calcium Efflux From Awake Cat Cerebral Cortex." *Bioelectromagnetics* 3 (1982): 295–308.

Alexander, Gordon, and Alexander, Douglas. *Biology*, 9th ed. New York: Barnes & Noble, Inc., 1971.

Baker, Jeffrey J. W., and Allen, Garland E. *The Study of Biology*. Reading, MA: Addison-Wesley, 1967.

Baker, R. R. *Magnetite Biomineralization and Magnetoreception in Animals: A New Biomagnetism*, eds. J. L. Kirschvink, D. S. Jones, and B. J. MacFadden. New York: Plenum Press, 1985, pp. 537–62.

Baker, R. R., Mather, J. G., and Kennaugh, J. H. *Nature* 301 (1983): 78–80.

Bawin, S., and Adey, W. R. "Sensitivity of Calcium Binding in Cerebral Tissue to Weak Environmental Electric Fields Oscillating at Low Frequency." *Annals of New York Academy of Sciences* 247 (1976): 74.

Becker, Robert O. *Cross Currents: The Perils of Electropollution, the Promise of Electromedicine*. Los Angeles: Jeremy Tarcher, Inc., 1990.

Becker, Robert O., and Marino, Andrew. *Electromagnetism and Life*. Albany: State University of New York Press, 1982.

Becker, Robert O., and Selden, Gary. *The Body Electric*. New York: William Morrow, 1985.

Blackman, Carl F. "Do Electromagnetic Fields Pose Health Problems?" Colloquium presentation, April 18, 1990. In *Frontier Perspectives*, vol. 2, no. 2, Fall/Winter 1991.

Blackman, Carl F., et al. "Calcium Efflux from Brain Tissue: Power Density Versus Internal Field Intensity Dependencies at 50 MHz RF Radiation." *Bioelectromagnetics* 1 (1989): 277–83.

Blackman, Carl F., et al. "Effects of ELF (1–120 Hz) and Modulated (50 Hz) RF Fields on the Efflux of Calcium Ions for Brain Tissue in Vitro." *Bioelectromagnetics* 6 (1985): 1–11.

Byrus, C. V., Pieper, S. E., and Adey, W. R. "The Effects of Low Energy 60 Hz Environmental Electromagnetic Fields upon the Growth-

Related Enzyme Ornithine Decarboxylase." *Carcinogenesis* 8 (1987): 1385–89.

———. "Increased Ornithine Decarboxylase Activity in Cultured Cells Exposed to Low Energy Modulated Microwave Fields and Phorebol Ester Tumor Promoters." *Cancer Research* 48 (1988): 4222–26.

Conference Notes from Transmission & Distribution Magazine's EMF Conference, 1993. Paper presented by Joseph Kirschvink.

Davis, D. L., Dinse, G. E., and Hoel, D. G. "Decreasing Cardiovascular Disease and Increasing Cancer Among Whites in the United States from 1973 Through 1987." *Journal of the American Medical Association,* vol. 271, no. 6, Feb. 9, 1994, pp. 431–37.

Kirschvink, Joseph L. *Magnetite Biomineralization in the Human Brain.* Pasadena, CA: Caltech Geological and Planetary Dept., 1992.

Nordenström, B. "Biologically Closed Electric Circuits: Activation of Vascular, Interstitial, Closed, Electrical Circuits for Treatment of Inoperable Cancers." *Journal of Bioelectricity* 3:162 (1984): pp. 137–53.

———. *Biologically Closed Electric Circuits: Clinical, Experimental, and Theoretical Evidence for an Additional Circulatory System.* Stockholm: Nordic Medical, 1983.

Chapter 9

Barfub, H., et al. "Whole-Body MR Spectroscopy of Humans at 4.0 Tesla." *Society for Magnetic Resonance in Medicine Conference Proceedings,* 1989–1990.

Becker, Robert O. *Cross Currents: The Perils of Electropollution, the Promise of Electromedicine.* Los Angeles: Jeremy Tarcher, Inc., 1990.

Becker, Robert O., and Selden, Gary. *The Body Electric.* New York: William Morrow, 1985.

Brodeur, Paul. *Currents of Death: Power Lines, Computer Terminals, and the Attempt to Cover Up Their Threat to Your Health.* New York: Simon & Schuster, 1989.

Chiles, C., et al. "Effect of Magnetic Fields on Nicotinic Acetylcholine Receptor Function." *Society for Magnetic Resonance in Medicine Conference Proceedings,* 1989–1990.

Delgado, J. N., Leal, J., et al. "Embryonic Changes Induced by Weak Extra-Low Frequency EMF's." *Journal of Anatomy* 134 (1982): 533–51.

Frazier, Marvin E. "Biologic Effect of Magnetic Fields: A Progress Report." In *Health Implications of New Energy Technologies,* ed. W. Rom and V. Archer. Ann Arbor, MI: Science Publishing Group.

Fitzgerald, Karen. "Magnetic Apprehensions: Radiologists Call for More Testing of MRI's Effects." *Scientific American,* Oct. 1993.

———. "New MRI Technologies Raise Safety Concerns." *IEEE Spectrum,* vol. 15, no. 6, July/Aug. 1991.

Goldhaber, G., Polen, M., and Hiatt, R. "Risk of Miscarriage and Birth Defects Among Women Who Use Video Display Terminals During Pregnancy." *American Journal of Industrial Medicine,* 13 (1988): 695–706.

Goodman, Reba. *Proceedings of the National Academy of Sciences* 85 (1988): 3298.

Goodman, R., and Henderson, A. "Transcriptional Patterns in X Chromosome of Sciara Following Exposure to Magnetic Fields." *Bioelectromagnetics* 8 (1987): 1–7.

Hoult, D. I., and Chen, C. N. "The Visualization of Probe Electric Fields Reveals Intense Hot-spots." *Society for Magnetic Resonance in Medicine Conference Proceedings,* 1989–1990.

Manikowska-Czerska, E., Czerska, P., and Leach, W. M. *Proceedings of U.S.-USSR Workshop on Physical Factors—Microwaves and Low Frequency Fields.* Washington, D.C.: National Institute of Environmental Health Sciences, 1985.

Marino, Andrew A., and Becker, Robert O. "The Effect of Continuous Exposure to Low Frequency Electric Fields on Three Generations of Mice." *Experientia* 32 (1976): 505–7.

———. "Power Frequency Electric Field Induced Biological Changes in Successive Generations of Mice." *Experientia* 35 (1980): 309–11.

Martin, A. H. "Magnetic Fields and Time Dependent Effects on Development." *Journal of Bioelectromagnetics* 9 (1988): 393–96.

Microwave News, Nov./Dec. 1992. "Danish Studies Offer New Support for EMF-Cancer Link," p. 5.

———, Nov./Dec., 1994. "Fetal Loss Found Among Rats Exposed to DC Magnetic Fields," p. 11.

"Natural Electrical Fields in Embryos." *EMF Health Report,* vol. 1, no. 1, June 1993. (From *Development* 114:4 [1992]: 985–96, describing research at Purdue University.)

Nordenson, I., et al. *Radiation and Environmental Biophysics* 23 (1984): 191.

Nordstrom, S., et al. "Effects of Paternal EMF Exposure on Offspring." *Bioelectromagnetics* 4 (1983): 91–101.

Ossenkopp, Klaus-Peter, et al. "Exposure to Nuclear Magnetic Resonance Imaging Procedure Attentuates Morphine-Induced Analgesia in Mice." *Life Sciences*, vol. 37, pp. 1507–14. Pergamon Press, 1985.

"Parkinson's Disease: One Day at a Time." *Harvard Health Letter*, vol. 18, no. 8, June 1993.

Phillips, Richard. "Biological Effects of Electric Fields on Miniature Pigs." *Proceedings of the Fourth Workshop of the U.S.-USSR Scientific Exchange on Physical Factors in the Environment*. National Institute of Environmental Health Science, June 21–24, 1983.

Prasad, N., et al. "Effects of MR Imaging on Murine Natural Killer Cell Cytoxicity." *American Journal of Radiology* 148(2), Feb. 1987, p. 415.

Redington, R. W., et al. "MR Imaging and Bio-Effects in a Whole Body 4.0 Tesla Imaging System." Presented at New York Academy of Sciences Conference on Biological Effects and Safety Aspects of Nuclear Magnetic Resonance Imaging and Spectroscopy, May 15–17, 1991.

Salzinger, Kurt. *Biological Effects of Power Line Fields*. Albany, NY: New York State Power-Lines Project Scientific Advisory Panel, 1987.

Shellock, F. G., et al. "Effect of a 1.5 Tesla Static Magnetic Field on the Body and Skin Temperature of Man." *Society for Magnetic Resonance in Medicine Conference Proceedings*, 1989–1990.

Simpson, Joe Leigh; Globus, Mitchell S.; Martin, Alice; and Sarto, Gloria. *Genetics in Obstetrics and Gynecology*. New York: Grune & Stratton, Inc., 1982.

Smith, Cyril, and Best, Simon. *Electromagnetic Man*. New York: St. Martin's Press, 1989.

Stolwijk, Jan A. J. *Electromagnetic Field Health Effects: Response to Inquiry*. State of Connecticut: Department of Health Services, Connecticut Academy of Science and Engineering, 1992.

Sulzman, Frank. *Biological Effects of Power Line Fields*. Albany, NY: New York State Power-Lines Project Scientific Advisory Panel, 1987.

Sussman, John R., and Levitt, B. Blake. *Before You Conceive: The Complete Prepregnancy Guide.* New York: Bantam, 1989.
Tapley, Donald F., and Todd, W. Duane. *The Columbia University College of Physicians and Surgeons Complete Guide to Pregnancy.* New York: Crown Publishers, Inc., 1988.
U.S. Congress, Office of Technology Assessment. *Reproductive Health Hazards in the Workplace.* Washington, D.C.: U.S. Government Printing Office, OTA-BA-266, Dec. 1985.
Wertheimer, N., and Leeper, E. "Possible Effects of Electric Blankets and Heated Waterbeds on Fetal Development." *Bioelectromagnetics* 7 (1986): 13–22.

Chapter 10

Altman, Lawrence K. "Chronic Fatigue Syndrome Finally Gets Some Respect." *New York Times,* Dec. 4, 1990.
———. "U.S. Scientists Baffled by a Cuban Epidemic." *New York Times,* May 30, 1993.
———. "Virus Seen as Possible Link to Chronic Fatigue Ailment." *New York Times,* Sept. 5, 1990.
"Alzheimer's in the Skin." *Discover,* Dec. 1993, p. 26.
Angier, Natalie. "New Clues Found in Fatal Illnesses." *New York Times,* June 6, 1993.
"Anti-Inflammatory Drugs May Impede Alzheimer's." *New York Times,* Feb. 20, 1994, p. 27.
Becker, Robert O. *Cross Currents: The Perils of Electropollution, the Promise of Electromedicine.* Los Angeles: Jeremy P. Tarcher, Inc., 1990.
Becker, Robert O., and Selden, Gary. *The Body Electric.* New York: William Morrow, 1985.
Brodeur, Paul. *Currents of Death: Power Lines, Computer Terminals, and the Attempt to Cover Up Their Threat to Your Health.* New York: Simon & Schuster, 1989.
Castleman, Michael. "Beach Bummer." *Mother Jones,* May/June 1993, p. 33.
———. "Breast Cancer and the Environment." *Mother Jones,* May/June 1994, p. 34.

Conference Notes from Transmission & Distribution Magazine's EMF Conference, 1992. Paper presented by David O. Carpenter.

Conference Notes from Transmission & Distribution Magazine's EMF Conference, 1993. Papers presented by Imre Gyuk, Larry Anderson, Birgitta Floderus, and Anders Ahlbom.

Conference Notes from Transmission & Distribution Magazine's EMF Conference, 1994. Paper presented by Eugene Sobel.

Ehrlich, Gretel. "I Have Been Struck by Lightning and I Am Alive." *Health,* May/June 1994.

"Electronic Blizzard Brings Down U.S. Planes." *Monitoring Times,* May 1990.

"Health Panel Finds No Single Cause for Gulf Veterans' Illnesses." *New York Times,* May 1, 1994.

Henig, Robin Marantz. "Is Misplacing Your Glasses Alzheimer's?" *New York Times Magazine,* April 24, 1994.

"Lou Gehrig's Disease: More Questions Than Answers," *Harvard Health Letter,* vol. 20, no. 4, Feb. 1995, p. 6.

Microwave News, Nov./Dec. 1985: "Fluorescent Lights and Skin Cancer," p. 12.

————, Jan./Feb. 1987: "Will Empress II Close Down Baltimore Harbor?" p. 2.

————, March/April 1990: "Are Brain Tumors Markers for EMF Exposure?" p. 1; "RF/MF Stimulates and Suppresses Human Brain Tumor Cells," p. 5; "EMF Brain Tumor Studies," p. 15.

————, Sept./Oct. 1991: "Breast Cancer and EMF's: Recent Papers," p. 3; "Five More Cancer-Police Radar Claims Filed," p. 12; "Black Hawk Helicopter Accidents," p. 13.

————, May/June 1992: "Human and Cellular Studies Point to Similar Mutagenic Effects of Radar," p. 1; "Navy Shelves Plans for Gulf Coast Empress II Site," p. 9; "Gulf War EMP Weapon Reported," p. 10.

————, July/Aug. 1992: "Male Breast Cancer Among Workers Under 65," p. 8; "Australian Study Finds CRT-Brain Tumor Risk," p. 9; "Melanoma Linked to Fluorescent Lights," p. 10.

————, Sept./Oct. 1992: "Swedish Officials Acknowledge EMF-Cancer Connection," p. 1; "Are Navy Communications Towers Responsible for Hawaiian Childhood Leukemia Cluster?" p. 5; "Similar Cluster Near U.S. Navy Transmitters in Scotland," p. 5; "Norwegian and Danish Studies Also Point to Cancer Risk," p. 13; "Abstracts of the Swedish Epidemiological Studies," p. 14.

———, Nov./Dec. 1992: "A Conversation with Dr. Anders Ahlbom," p. 3; "Danish Studies Offer New Support of EMF-Cancer Link," p. 5; "Brain Tumor Recognized As Work-Related Injury in Sweden," p. 5.

———, Jan./Feb. 1993: "Cancer Excess at Aluminum Plant in Washington State," p. 3.

———, March/April 1993: "Acute Leukemia Linked to EMF Exposure," p. 7; "Leukemia Among Telephone Linemen," p. 7; "NHL Among Swedish Power Workers," p. 7; "Navy to Cancel Empress II," p. 14; "VDT Research Center Weighs EMF-Breast Cancer Studies," p. 14.

———, May/June 1993: "NAS-NRC Finds GWEN Poses Minimal Public Health Risk," p. 11; "Florida ALS Cluster," p. 16.

———, Sept./Oct. 1993: "Police Radar Injury Litigation After Bendure: An Update," p. 8; "ALS Cluster Prompts Study, Meeting," p. 16.

———, Nov./Dec. 1993: "Breast Cancer Risk Found for Female Electrical Workers," p. 1; "Pooled Nordic Data Support Childhood Leukemia Risk," p. 5.

———, Jan./Feb. 1994: "Florida Lawsuit Blames Couple's Rare Leukemia on EMF's," p. 1; "MAGLEV, Effects of Simulated Fields," p. 15.

———, March/April 1994: "Canadian-French Study Supports Occupational Leukemia Risk, Weaker EMF Link to Brain Tumors," p. 1.

Phillips, J. L., and Winters, W. D. "In Vitro Exposure to EMF's: Changes in Tumor Cell Properties." *International Journal of Radiation Biology* 46 (1986): 463–69.

———. "Transferring Binding to Two Human Colon Carcinoma Cell Lines: Characterizations and Effect of 60-Hz EMF's." *Cancer Research* 46 (1986): 239–44.

Randegger, E. "Electromagetic Pollution." *Environ* no. 7.

Simons, Marlise. "Dead Mediterranean Dolphins Give Nations Pause." *New York Times*, Feb. 2, 1992.

Smith, Cyril W., and Best, Simon. *Electromagnetic Man*. New York: St. Martin's Press, 1989.

Stoff, Jesse A., and Pellegrino, Charles R. *Chronic Fatigue Syndrome, the Hidden Epidemic*. New York: Harper and Row, 1990.

U.S. Environmental Protection Agency. *Evaluation of the Potential*

Carcinogenicity of Electromagnetic Fields. Draft review, EPA/600/6-90/005B, Oct. 1990.

Wertheimer, Linda. "Are Memory and Hormones Related?" National Public Radio, "All Things Considered," Nov. 9, 1993.

"Wives of Sick Gulf War Vets Also Reporting Illnesses." Danbury, *The News-Times,* March 9, 1993.

Chapter 11

Becker, Robert O. *Cross Currents: The Perils of Electropollution, the Promise of Electromedicine.* Los Angeles: Jeremy P. Tarcher, Inc., 1990, p. 283.

Electric and Magnetic Fields from 60 Hertz Electrical Power: What Do We Know About Possible Health Risks? Pittsburgh Department of Engineering and Public Policy, Carnegie Mellon University, 1989.

"Electromagnetic Fields." *Consumer Reports,* May 1994, pp. 354–59.

"EMF, How Dangerous?" *Transmission & Distribution,* ed. E. Hazan, E. G. Enabnit, Jr., and B. H. LeCerf, June 1991, pp. 15–30.

"EPA Lab Tests EFM131, EFM140—Finds Highly Accurate." *EMF Company Newsletter,* no. 4, Dec. 1992.

MacIntyre, Steven A. *Tutorial, ELF Magnetic Field Measurement Methods and Accuracy Issues.* Meda, Inc.

Microwave News, March/April 1993, p. 8; May/June 1994, pp. 3–4, 10.

Riley, Karl. *A Frank Talk About Measurement and Measurement Programs.* Tucson, AZ: Magnetic Sciences International.

———. *MSI Magnetic Field Guide: A General Introduction to the Measurement of Magnetic Fields in Home and Work Environments,* 2nd rev. ed. Tucson, AZ: Magnetic Sciences International, 1993.

———. *Tracing EMFs in Building Wiring and Grounding.* Tucson, AZ: Magnetic Sciences International, 1995.

Chapter 12

Becker, Robert, O. *Cross Currents: The Perils of Electropollution, the Promise of Electromedicine.* Los Angeles: Jeremy P. Tarcher, Inc., 1990, pp. 267–81.

Blanc, Bernard H., and Hertel, Hans U. "Influence on Man: Comparative Study About Food Prepared Conventionally and in the Microwave Oven." *Raum & Zeit,* vol. 3, no. 2, 1992.

Chen, J.-Y., and Gandhi, O. P. "RF Currents Induced in an Anatomically-Based Model of a Human for Place-Wave Exposures (20–100 MHz)." *Health Physics* 57 (July 1989): 89–98.

Cleveland, R. F., and Athey, T. W. "Specific Absorption Rate (SAR) in Models of the Human Head Exposed to Hand-Held UHF Portable Radios." *Bioelectromagnetics* 10 (1989): 173–86.

Elder, J. A., Czerski, M. A., et al. *The Effects of Non-Ionizing Radiation, Radiofrequency Radiation.* World Health Organization, 1989.

"EMF, How Dangerous?" *Transmission & Distribution,* ed. E. Hazan, E. G. Enabnit, Jr., and B. H. LeCerf, June 1991, pp. 15–30.

Hartwell, Fred. "Magnetic Fields from Water Pipes." *EC&M,* vol. 92, no. 3, March 1993, pp. 63–70.

Keller, John J. "Cellular Phone Safety Concerns Hammer Stocks." *Wall Street Journal,* Jan. 25, 1993.

———. "McCaw to Study Cellular Phones As Safety Questions Affect Sales." *Wall Street Journal,* Jan. 29, 1993.

Macaulay, David. *How Things Work.* Boston: Houghton Mifflin, 1988.

Microwave News, Nov./Dec. 1985: "Fluorescent Lights and Skin Cancer," p. 12.

———. May/June 1992: "Microwaving Breast Milk," p. 14.

———. July/Aug. 1992: "Melanoma Linked to Fluorescent Lights," p. 10.

———. Jan./Feb. 1994: "No Consensus on Cellular Phone RF Radiation Levels in Brain," p. 1.

Riley, Karl. *Tracing EMFs in Building Wiring and Grounding.* Tucson, AZ: Magnetic Sciences International, 1994.

U.S. Environmental Protection Agency. *EMF in Your Environment: Magnetic Field Measurements of Everyday Electrical Devices.* EPA Office of Radiation and Indoor Air, Doc. 402-R-92-008, Dec. 1992.

White, D.; Barge, J. M.; George, E. A.; and Riley, K. *The EMF Controversy and Reducing Exposure from Magnetic Fields.* Gainesville, VA: Interference Control Technologies, Inc., 1993.

Chapter 13

Adey, William Ross. Testimony before U.S. Senate, Ad Hoc Subcommittee on Consumer and Environmental Affairs. Sen. Joseph I. Lieberman, chairman, "Hearing on Health Risks Posed by Radar Guns:

The Extent of Federal Research and Regulatory Development of Microwave Emissions From Hand-Held Radar Guns," Aug. 7, 1992.

Balcer-Kubczek, E. K., and Harrison, G. H. "Evidence for Microwave Carcinogenesis in Vitro." *Carcinogenesis* 6 (1989): 859–64.

Becker, Robert O. *Cross Currents: The Perils of Electropollution, the Promise of Electromedicine*. Los Angeles: Jeremy P. Tarcher, Inc., 1990, pp. 267–81.

Becker, Robert O., and Selden, Gary. *The Body Electric*. New York: William Morrow, 1985, pp. 271–317.

Bowman, J., et al. "Electric and Magnetic Field Exposure, Chemical Exposure, and Leukemia Risk in 'Electrical' Occupations." EPRI Report TR-101723, Dec. 1992.

Bowman, J. D., et al. "Exposures to Extra Low Frequency EMF's in Occupations with Elevated Leukemia Rates." *Applied Industrial Hygiene* 36 (1988): 189–93.

Breysse, P., et al. "ELF Magnetic Field Exposures in an Office Environment." *American Journal of Industrial Medicine* 25 (Feb. 1994): 177–85.

Chang, B. K., et al. "Inhibition of DNA Synthesis by Microwave Radiation in L1210 Leukemia Cells." *Cancer Research* 40 (1980): 1002–5.

Coleman, M., and Berel, V. "Review of Epidemiological Studies of the Health Effects of Living or Working with Electricity Generation or Transmission Equipment." *International Journal of Epidemiology* 17 (1988): 1–13.

Floderus, B., et al. *Occupational Exposure to Electromagnetic Fields in Relation to Leukemia and Brain Tumors: A Case-Control Study*. Solna, Sweden: Department of Neuromedicine at the National Institute of Occupational Health, 1993.

Flodin, U., et al. "Background Radiation, Electrical Work, and Some Other Exposures Associated with Acute Myeloid Leukemia." *Archives of Environmental Health* 41:2 (1986): 77–84.

Frey, A. H., et al. "Neural Function and Behavior." *Annals of New York Academy of Science* 247 (1975): 433–38.

Garland, E., et al. "Incidence of Leukemia in Occupations with Potential EMGF Exposure in U.S. Navy Personnel." *American Journal of Epidemiology* 132:10 (1991): 293.

Gilman, P. A. "Leukemia Risk Among U.S. White Male Coal Miners." *Journal of Occupational Medicine* 27:9 (1985): 669–71.

Guy, A. W., et al. "Effects of Long-Term Low-Level RF Radiation Exposure on Rats." Vol. 9, University of Washington. USAFSAM-TR-85, Aug. 1985.

Howe, G. R., and Lindsay, J. P. "Cancer Mortality in Males." *Journal of the National Cancer Institute* 70 (1983): 37–44.

Juutilainen, J., et al. "Results of an Epidemiological Cancer Study Among Electrical Workers in Finland." *Journal of Bioelectromagnetism* 7:1 (1988): 119–21.

Lai, H. and Singh, N. "Acute Low-Intensity Microwave Exposure Increases DNA Single-Strand Breaks in Rat Brain Cells," *Bioelectromagnetics* 16, 1995.

Lester, J. R., and Moore, D. F. "Airport Radar and Incidence of Cancer in Wichita." *Journal of Bioelectricity* 1:1 (1984).

Lin, R., et al. "Occupational Exposure to EMF's and the Occurrence of Brain Tumors." *Journal of Occupational Medicine* 27:6 (1985): 413–19.

Linet, M. S., et al. "Leukemia and Occupation in Sweden." *American Journal of Industrial Medicine* 14 (1988): 319–30.

London, S. J., et al. *American Journal of Epidemiology* 134 (1991): 923.

Loomis, D. P., and Savitz, D. A. "Brain Cancer and Leukemia Mortality Among Electrical Workers." *American Journal of Epidemiology* 130 (1989): 814.

Mack, C., et al. "Astrocytoma Risk Related to Job Exposure to Electric and Magnetic Fields." *Bioelectromagnetics* 12 (1991): 57–66.

Mantanowski, G., et al. "Leukemia in Telephone Company Employees." *Contractors Review*, U.S. Department of Energy/EPRI, 1988.

Mantanowski, G., Elliot, E., and Breysse, P. "Cancer Incidence in New York Telephone Workers." *Annual EPRI Contractors Review*, Nov. 15, 1989.

McDowall, M. N., et al. "Leukemia Mortality in Electrical Workers in England and Wales." *Lancet*, 1983, p. 246.

Microwave News, Jan./Feb. 1987: "New Support for Cancer Risk from RF/MW Exposure," p. 1; "MWs As Co-Teratogens," p. 1.

———, Sept./Oct. 1989: "AM/FM Radiation Litigation: Two Cancer Settlements Are Secret, Two Other Suits Are Pending," p. 1.

———, March/April 1990: "Are Brain Tumors Markers for EMF Exposure?" p. 1; "RF/MW Stimulates and Suppresses Human Brain Tumor Cells," p. 5.

———, Sept./Oct. 1991: "Microwaves Cause Persistent Eye Damage Below 4 W/Kg," p. 1; "Demers Male Breast Cancer Study Published," p. 3.

———, May/June 1992: "Human and Cellular Studies Point to Similar Mutagenic Effect of Radar," p. 1.

———, July/Aug. 1992: "Male Breast Cancer Among Workers Under 65," p. 8; "Australian Study Finds CRT-Brain Tumor Risk," p. 9.

———, Sept./Oct. 1992: "Are Navy Communications Towers Responsible for Hawaiian Childhood Leukemia Cluster?" p. 5; "Similar Cluster Near U.S. Navy Transmitters in Scotland," p. 5.

———, Jan./Feb. 1993: "Cancer Excess at Aluminum Plant in Washington State," p. 3.

———, March/April 1993: "EMF Mitigation: Cost-Effective in New Buildings, Costly in Existing Offices," p. 1; "Acute Leukemia Linked to EMF Exposure," p. 7; "Leukemia Among Telephone Linemen," p. 7; "NHL Among Swedish Power Workers," p. 7; "EMF's at New York City's Best Addresses," p. 11; "EMS Technicians Question Safety of Their Two-Way Radios," p. 13.

———, May/June 1993: "EMP, Effects on Medical Devices," p. 15; "Radar, Florida ALS Cluster," p. 16.

———, Sept./Oct. 1993: "Police Radar Injury Litigation After Bendure: An Update," p. 8; "Radar, Phased Array v. Scanning," p. 16.

———, Nov./Dec. 1993: "Breast Cancer Risk Found for Female Electrical Workers," p. 1.

———, Jan./Feb. 1994: "Miscarriage Risk for Therapists Using Microwave Diathermy," p. 11; "MAGLEV, Effects of Simulated Fields," p. 15; "MRI, Miscarriage Risks Among Technicians," p. 17.

———, March/April 1994: "Canadian-French Study Supports Occupational Leukemia Risk, Weaker EMF Link to Brain Tumors," p. 1; "Transient EMF's May Be a Key to the Mechanism of Interaction," p. 1.

———, May/June 1994: "New Clues on Leukemia and Breast Cancer Risks Among Railway Workers," p. 1; "Swedish Trade Union Group Demands EMF Safeguards," p. 5; "The Pitfalls of Using Job Titles to

Assess EMF Exposures," p. 9; "Police Radar Class Action Suit Seeks Medical Monitoring," p. 11.

————, July/Aug. 1994: "Radar and Office EMF's Suspected in Census Bureau Brain Tumor Cluster," p. 1; "Digital Cellular Phones Can Disrupt Implanted Pacemakers," p. 1; "Alzheimer's Disease Linked to EMF's," p. 4; "Breast Cancer Risk for Female Radio Operators in Norway," p. 5; "USC Study Finds Small Leukemia Risk for Electrical Workers," p. 7.

————, Nov./Dec. 1994: "Microwaves Break DNA in Brain; Cellular Phone Industry Skeptical," p. 1.

Milham, S. "Mortality from Leukemia in Workers Exposed to Electric and Magnetic Fields." *New England Journal of Medicine* 307 (1982): 249.

————. "Silent Keys: Leukemia Mortality in Amateur Radio Operators." *Lancet,* April 6, 1985, p. 812.

Nordenson, I., et al. "Effects in Human Lymphocytes of Power Frequency Electric Fields." *Radiation and Environmental Biophysics* 23 (1984): 191–201.

Nordstrom, S., et al. "Effects of Paternal EMF Exposure on Offspring." *Bioelectromagnetics* 4 (1983): 91–101.

Pearce, N. E., et al. "Leukemia in Electrical Workers in New Zealand." *Lancet,* April 1, 1985, pp. 811–12.

Reif, J. S., et al. "Occupational Risks for Brain Cancer." *Journal of Occupational Medicine* 31:10 (1989): 863.

Sandstrom, M. "A Survey of Electric and Magnetic Fields Among VDT Operators in Offices." *IEEE Transaction on Electromagnetic Compatibility* 35, Aug. 1993, pp. 394–97.

Sarkar, S., Ali, S. and Behari, J. "Effect of Low-Power Microwave on the Mouse Genome: A Direct DNA Analysis," *Mutation Research,* 320, pp. 141–47, 1994.

Savitz, D. A., and Calle, E. E. "Leukemia and Occupational Exposure to EMF's: Review of the Epidemiologic Surveys." *Journal of Occupational Medicine* 29:1 (1987).

Sobel, E. Paper on Alzheimer's and occupational EMF exposures. Presented at 4th International Conference on Alzheimer's Disease and Related Disorders, Minneapolis, Iowa. July 31, 1994.

Speers, M. A., et al. "Occupational Exposures and Brain Cancer

Mortality." *American Journal of Industrial Medicine* 13 (1988): 629–38.

Spitz, M., and Johnson, C. "Neuroblastoma and Paternal Occupation." *American Journal of Epidemiology* 121 (1985): 924–29.

Stodolnik-Baranska, W., et al. "Effects of Microwaves on Human Lymphocyte Cultures." *Proceedings of the International Symposium*, Warsaw, 1973, pp. 189–95.

Sugarman, Ellen. *Warning: The Electricity Around You May Be Hazardous to Your Health.* New York: Simon & Schuster, 1992, appendix A.

Szmigielski, S., et al. "Immunological and Cancer-Related Effects of Exposure to Low-Level Microwave and RF Fields." *Modern Bioelectricity*, ed. A. Marino. New York: Marcel Decker, 1988.

Takashima, S., et al. "Effects of Modulated RF Energy on the EEG's of Mammalian Brains." *Radiation and Environmental Biophysics* 16 (1979): 15–27.

Theriault, G., et al. "Cancer Risks Associated with Occupational Exposure to Magnetic Fields Among Electric Utility Workers in Ontario and Quebec, Canada, and France: 1970–1989." *American Journal of Epidemiology* 139 (1994): 550–72.

Thomas, T. L. "Brain Tumor Mortality Risk Among Men with Electrical and Electronic Jobs." *Journal of the National Cancer Institute* 79:2 (1987): 233–38.

Tornqvist, S., et al. "Cancer in the Electric Power Industry." *British Journal of Industrial Medicine* 43 (1986): 212–13.

Vagero, D., et al. "Cancer Morbidity Among Workers in the Telecommunications Industry." *British Journal of Industrial Medicine* 42 (1985): 191–98.

VDT News, March/April 1993: "Studies of EMF-Breast Cancer Link Proposed," p. 1.

———, May/June 1993: "Designing a Low EMF Office," p. 1; "EPA's Power-Down Initiative Gains Momentum," p. 7; "Demand Is Growing for Shielding Strong EMF's in Offices," p. 9.

———, Jan./Feb. 1994: "EMF-Miscarriage Link Supported," p. 1.

———, March/April 1994: "Swedish Union Comes to U.S. to Promote 'Green' Computing," p. 1; "EMF's in the Office: Looking Beyond VDTs," p. 4.

———, July/Aug. 1994: "Glare Screen Makers' Deceptive MPRII

Claims," p. 3; "Low EMF Monitor Watch," p. 10; "Breast Cancer Risk for Female Electrical Workers," p. 13.

Wilkins, J. R., et al. "Paternal Occupation and Brain Cancer in Offspring: A Mortality-Based Case-Control Study." *American Journal of Industrial Medicine* 14 (1988): 299–318.

Wright, et al. "Leukemia in Workers Exposed to Electric and Magnetic Fields." *Lancet* 1160 (1982): 61.

Yao, K. T. S. "Microwave Radiation Induced Chromosome Aberrations in Corneal Epithelium of Chinese Hamsters." *Journal of Heredity* 69 (1978): 409–12.

Chapter 14

Ahlbom, A., and Feychting, M. *Magnetic Fields and Cancer in People Residing Near Swedish High Voltage Power Lines.* Stockholm, Sweden: Institute of Environmental Medicine at Karolinska Institute, 1992.

Becker, Robert O. *Cross Currents: The Perils of Electropollution, the Promise of Electromedicine.* Los Angeles: Jeremy P. Tarcher, 1990, pp. 200–212.

Bowman, J. D. "The Risk of Childhood Leukemia from Home Exposure to Resonance from Static and Power Frequency Magnetic Fields." Abstract, U.S. Dept. of Energy Conference, Nov. 1991.

Carpenter, D., and Ayropetyan, S., eds. *Biological Effects of Electric and Magnetic Fields.* San Diego: Academic Press, 1994.

Conference Notes from Transmission & Distribution Magazine's EMF Conference, 1994. Papers presented by Eugene Sobel, Zoreh Davanipour, and David Savitz.

Electrical and Biological Effects of Transmission Lines. Portland, OR: U.S. Department of Energy, Bonneville Power Administration, 1989, pp. 11–22, 71–80.

Electric and Magnetic Fields from 60 Hertz Electric Power: What Do We Know About Possible Health Risks? Pittsburgh: Carnegie Mellon University, 1989, pp. 7, 11, 32.

Microwave News, Nov./Dec. 1985: "60-Hz Fields Induce Neurologic Changes in Monkeys," p. 4.

———, March/April 1988: "Henhouse Project: Weak PMF's Cause Chick Abnormalities," p. 1; "100 Hz Sine Waves Transform Cells," p. 4.

————, Sept./Oct. 1991: "Breast Cancer and EMF's: Recent Papers," p. 3; "USC EMF-Childhood Leukemia Study Due in November," p. 4.

————, Sept./Oct. 1992: "Swedish Officials Acknowledge EMF-Cancer Connection," p. 1.

————, Nov./Dec. 1992: "Danish Studies Offer New Support of EMF-Cancer Link," p. 5; "Brain Tumor Recognized As Work-Related Injury in Sweden," p. 5.

————, March/April 1993: "EMF Mitigation: Congress, State Seek to Limit Power Line EMFs at Schools," p. 4.

————, Sept./Oct. 1993: "Finnish Childhood Cancer Study: No Link Reported; Others Unsure," p. 4; "The Effect of EMF's on Property Value: Four Noteworthy Cases," p. 6.

————, March/April 1994: "Canadian-French Study Supports Occupational Leukemia Risk, Weaker EMF Link to Brain Tumors," p. 1; "School EMF's Alarm Parents and Teachers Across the Northeast," p. 6.

Peters, J. "Childhood Leukemia and Exposure to Electricity." *American Journal of Epidemiology* (Nov. 1991): 215–30.

Savitz, D. A., et al. "Case-Control Study of Childhood Cancer and Exposure to 60-Hz Magnetic Fields." *American Journal of Epidemiology* 128 (1989): 21–38.

Savitz, D. A., et al. "Magnetic Field Exposure from Electrical Appliances and Childhood Cancer." *American Journal of Epidemiology* 131 (1990): 763–73.

Savitz, D. A., and Kaune, W. "Childhood Cancer in Relation to a Modified Residential Wire Code." *Environmental Health Perspectives* 101, April 22, 1993, pp. 76–80.

Stern, F. B. "A Case-Control Study of Leukemia at a Naval Shipyard." *American Journal of Epidemiology* 123:6 (1986): 980–92.

Sugarman, Ellen. *Warning: The Electricity Around You May Be Hazardous to Your Health.* New York: Simon & Schuster, 1992, pp. 36–40, 120–27.

Tomenius, L. "50-Hz Electromagnetic Environment and the Incidence of Childhood Cancers in Stockholm County." *Bioelectromagnetics* 7 (1986): 191–207.

VDT News, Sept./Oct. 1994: "Alzheimer's Linked to EMF-Exposure," p. 1.

Wertheimer, N., and Leeper, E. "Adult Cancer Related to Electrical Wire Near the Home." *International Journal of Epidemiology* 11 (1982): 345–55.

———. "Electrical Wiring Configurations and Childhood Cancer." *American Journal of Epidemiology* 109 (1979): 273–84.

White, D.; Barge, J. M.; George, E. A.; and Riley, K. *The EMF Controversy and Reducing Exposure from Magnetic Fields.* 2.25–3.46. Gainesville, VA: Interference Control Technologies, Inc., 1993.

Wolpaw, J. R., et al. "Chronic Exposure of Primates to 60 Hz Electric and Magnetic Fields: I. Exposure System and Measurements of General Health and Performance; II. Neurochemical Effects; III. Neurophysiological Effects." *Bioelectromagnetics* 10 (1989): 277–319.

Chapter 15

Adey, William Ross. Testimony before U.S. Senate, Ad Hoc Subcommittee on Consumer and Environmental Affairs, Sen. Joseph I. Lieberman, chairman, "Hearing on Health Risks Posed by Radar Guns: The Extent of Federal Research and Regulatory Development of Microwave Emissions from Hand-Held Radar Guns," Aug. 7, 1992.

Baranski, S., and Czerski, P. *Biological Effects of Microwaves.* Stroudsburg, PA: Dowden Hutchingson and Ross, 1976.

Becker, Robert O. *Cross Currents: The Perils of Electropollution, the Promise of Electromedicine.* Los Angeles: Jeremy P. Tarcher, Inc., 1990.

Becker, Robert O., and Selden, Gary. *The Body Electric.* New York: William Morrow, 1985.

Benson, K. B. *Television Engineering Handbook.* New York: McGraw-Hill, 1986.

"Biological Effects and Exposure Criteria for Radiofrequency Electromagnetic Fields." Report No. 86, National Council on Radiation Protection and Measurements, 1986.

Bise, W. "Low Power Radio-Frequency and Microwave Effects on Human Electroencephalogram and Behavior." *Physiological Chemistry and Physics* 10 (1978): 387–97.

Cancer Incidence in Census Tracts with Broadcasting Towers in Honolulu, Hawaii. State of Hawaii: Dept. of Health, Environmental Epidemiology Program, Oct. 27, 1986.

Cleary, S., Liu, L.-M., and Merchant, R. *Radiation Research* 121 (1990): 38–45.

Elder, J. A., et al. *The Effects of Non-Ionizing Radiation.* World Health Organization, 1989.

"FCC Clearing Airwaves for Phones of the Future." *New York Times,* Sept. 20, 1993.

Frey, A. H. "Biological Function As Influenced by Low Power Modulated RF Energy." *IEEE Trans, Microwave Theory Tech.,* MTT-19, 153 (1971).

———. "Human Perception of Illumination with Pulsed Ultrahigh Frequency Electromagnetic Energy." *Science* 181 (1973): 356.

Frey, A. H., Feld, S. R., and Frey, B. "Neural Function and Behavior: Defining the Relationship." *Annals of New York Academy of Science* 247 (1975): 433.

Garaj-Vrhovac, V., Fucic, A., and Horvat, D. "Comparison of Chromosome Aberration and Micronucleus Induction in Human Lymphocytes After Occupational Exposure to Vinyl Chloride Monomer and Microwave Radiation." *Periodicum Biologorum* 92 (1990): 411–16.

———. "The Correlation Between the Frequency of Micronuclei and Specific Chromosome Aberrations in Human Lymphocytes Exposed to Microwave Radiation in Vitro." *Mutation Research* 281 (1992): 181–86.

Garaj-Vrhovac, V., Horvat, D., and Koren, Z. "The Effect of Microwave Radiation on the Cell Genome." *Mutation Research* 243 (1990): 87–93.

Goldoni, J. "Hematological Changes in Peripheral Blood of Workers Occupationally Exposed to Microwave Radiation." *Health Physics* 58 (1990): 205–7.

Goldoni, J., Durek, M., and Koren, Z. "Health Status of Personnel Occupationally Exposed to Microwaves and Radiofrequencies." *Proceedings of the 1st Congress of the European Bioelectromagnetics Association,* Jan. 23–25, 1992.

Gordon, Z. V., Roscin, A. V., and Byckov, M. S. "Main Directions and Results of Research Conducted in the U.S.S.R. on the Biologic Effects of Microwaves." *Biologic Effects and Health Hazards of Microwave Radiation.* Warsaw: Polish Medical Publishers, 1974, pp. 22–35.

Guy, A. W., et al. "Effects of Long-Term Low-Level Radiation Exposure on Rats." Vol. 9, University of Washington. USAFSAM-TR-85, Aug. 1985.

Henderson, A., et al. "Report: Effects of Broadcast Towers on Residential Cancer Rates." State of Hawaii: Dept. of Health, Environmental Epidemiology Program, 1986.

Investigation of Childhood Leukemia on Waianae Coast 1977–1990. State of Hawaii: Dept. of Health, Environmental Epidemiology Program, 1991.

Johnson, R. C., ed. *Antenna Engineering Handbook,* 3rd ed. 1993.

Klimkova-Deutschova, E. "Neurologic Findings in Persons Exposed to Microwaves." In *Biologic Effects and Health Hazards of Microwave Radiation.* Warsaw: Polish Medical Publishers, 1974, p. 270.

Lane, L. C. *Simplified Radiotelephone License Course,* vols. 1–3. New Jersey: Hayden Book Co., Inc., 1970.

Lester, J. R., and Moore, D. F. "Airport Radar and Incidence of Cancer in Wichita." *Journal of Bioelectricity* 1:1 (1984).

Lin, R. S., et al. "Occupational Exposure to Electromagnetic Fields and the Occurrence of Brain Tumors: An Analysis of Possible Associations." *Journal of Occupational Medicine* 27 (June 1985): 413–19.

Macaulay, David. *How Things Work.* Boston: Houghton Mifflin, 1988.

Marino, Andrew. Testimony in Friends of Fair Oaks Village . . . *v.* Pac Tel Cellular, Superior Court of California in Sacramento, No. 372242, Oct. 16, 1992.

———. Testimony in Eric Bendure *v.* Kustom Signals, Inc., U.S. District Court, Northern Dist. of California, Civil No. C9101173SAW, July 9, 1992.

Microwave News, Jan./Feb. 1989: "Vernon Down's Syndrome Study Abandoned; Critics Voice Outrage," p. 1.

———, Sept./Oct. 1989: "AM/FM Radiation Litigation: Two Cancer Settlements Are Secret, Two Other Suits Are Pending," p. 1; "Seattle Mayor Recommends a 200 μW/cm^2 Standard," p. 6.

———, March/April 1990: "Are Brain Tumors Markers for EMF Exposure?" p. 1; "RF/MW Stimulates and Suppresses Human Brain Tumor Cells," p. 5.

———, May/June 1992: "Cellular Telephone Radiation Blamed for Brain Tumor," p. 1; "Human and Cellular Studies Point to Similar Mutagenic Effects of Radar," p. 1; "Radar Radiation: A Call for Research," p. 12.

———, July/Aug. 1992: "Judge Extends Time for Filing RF/MW Radiation Exposure Claims," p. 1; "Puerto Rico Adopts RF/MW Rules," p. 15.

———, Sept./Oct. 1992: "Are Navy Communications Towers Responsible for Hawaiian Childhood Leukemia Cluster?" p. 5; "Similar Cluster Near U.S. Navy Transmitters in Scotland," p. 5; "BBC Tower Sparks Controversy," p. 11; "Radar and Power Line Fears in Ireland," p. 11.

———, May/June 1993: "The RF Problem," p. 13; "Florida ALS Cluster," p. 16.

———, Sept./Oct. 1993: "ANSI RF/MW Standard Challenged, U.S. Air Force and Hughes Units Adopt Limits Up to 100 Times Stricter," p. 1; "U.S. Air Force v. U. S. Air Force: Labs at Odds over RF/MW Health Risks," p. 1; "Biological Effects of Microwave Radiation: A White Paper," p. 12; "ALS Cluster Prompts Study, Meeting," p. 16; "Phased Array v. Scanning," p. 16.

———, Nov./Dec. 1993: "San Francisco Bans Cellular Antennas on School Property," p. 1; "Neurological Complaints near Swiss Shortwave Transmitter," p. 8; "Opposition Builds to Siting Cellular Towers on School Property," p. 11.

———, Jan./Feb. 1994: "EPA Assails ANSI RF/MW Standards As Seriously Flawed," p. 10.

———, March/April 1994: "Opposition to Weather Radar in California Stirs Controversy," p. 10; "Health Concerns Raised over Russian Radar in Latvia," p. 11; "RFI from Ham Radio Prompts Health Worries and Lawsuit," p. 12.

———, Sept./Oct. 1994: "EPA to Assess Health Impacts of Weak, Modulated RF/MW Radiation," p. 1; "Latvia's Russian Radar May Yield Clues to RF Health Risks," p. 12.

Milolajczyk, H. J. "Microwave Irradiation and Endocrine Functions." In *Biologic Effects and Health Hazards of Microwave Radiation*. Warsaw: Polish Medical Publishers, 1974, pp. 46–51.

Petersen, R. C., and Testagrossa, P. A. "Radio-Frequency Electromagnetic Fields Associated with Cellular-Radio Cell-Site Antennas." *Bioelectromagnetics* 13 (1992): 527–42.

"A Practical Guide to the Determination of Human Exposure to Radiofrequency Fields." Report No. 119, National Council on Radiation Protection and Measurements, 1993.

Sadcikova, M. N. "Clinical Manifestations of Reactions to Microwave Irradiation in Various Occupational Groups." In *Biologic Effects and Health Hazards of Microwave Radiation.* Warsaw: Polish Medical Publishers, 1974, pp. 261–63.

Sarkar, S., Ali, S., and Behari, J. *Effects of Low Power Microwaves on the Mouse Genome: A Direct DNA Analysis.* Delhi, India: Institute of Nuclear Medicine and Allied Sciences, Jan. 1994.

"Siting Cellular Transmitters." *Zoning News.* American Planning Association, Jan. 1991.

Smith, Cyril W., and Best, Simon. *Electromagnetic Man.* New York: St. Martin's Press, 1989.

Stein, D. "EMF Fear: Can It Tear Your Tower Down?" *Television Broadcast,* July 1994, p. 9.

Sugarman, Ellen. *Warning: The Electricity Around You May Be Hazardous to Your Health.* New York: Simon & Schuster, 1992.

Szmigielski, S., et al. "Immunological and Cancer-Related Effects of Exposure to Low-Level Microwave and RF Fields." *Modern Bioelectricity,* ed. Andrew Marino. New York: Marcel Decker, 1987.

Tell, R. A., and Mantiply, E. D. "Population Exposure to VHF and UHF Broadcast Radiation in the United States." *Proceedings of IEEE,* vol. 68, no. 1, Jan. 1980, pp. 6–12.

U.S. Environmental Protection Agency. *Evaluation of the Potential Carcinogenicity of Electromagnetic Fields.* Draft Review, EPA/600/6-90/005B, Oct. 1990.

U.S. Environmental Protection Agency. "Federal Radiation Protection Guidance: Proposed Alternatives for Controlling Public Exposure to Radiofrequency Radiation; Notice of Proposed Recommendations." *Federal Register,* vol. 51, no. 146, July 30, 1986.

U.S. Federal Communications Commission. "Responsibility of the Federal Communications Commission to Consider Biological Effects of Radiofrequency Radiation When Authorizing the Use of Radiofrequency Devices." Record, vol. 7, Doc. FCC 87-63, Gen. Docket No. 79-144. Adopted Feb. 12, 1987.

U.S. Federal Communications Commission. "Questions and Answers About Biological Effects and Potential Hazards of Radiofrequency Radiation." Office of Engineering and Technology, Bulletin No. 56, 3rd ed., 1989.

INDEX

419

978-0-595-47607-7
0-595-47607-4

Made in the USA
Lexington, KY
23 June 2012